THE QUOTABLE ATHEIST

Ammunition for
nonbelievers,
political junkies,
gadflies,
and those generally hell-bound

Jack Huberman

NATION BOOKS • NEW YORK

To Marvin,
with whom I most respectfully and lovingly disagree,
and who remains my all-knowing big brother.

THE QUOTABLE ATHEIST:
Ammunition for Nonbelievers, Political Junkies, Gadflies,
and Those Generally Hell-Bound

Published by
Nation Books
An Imprint of Avalon Publishing Group, Inc.
245 West 17th St., 11th Floor
New York, NY 10011

AVALON
publishing group incorporated

Copyright © 2007 by Jack Huberman

Nation Books is a copublishing venture of The Nation Institute and
Avalon Publishing Group, Incorporated.

Library of Congress Cataloging-in-Publication Data is available.

ISBN-10: 1-56025-969-8
ISBN-13: 978-1-56025-969-5

9 8 7 6 5 4 3 2

Book design by Maria Fernandez

Printed in the United States of America
Distributed by Publishers Group West

CONTENTS

INTRODUCTION

"The supernatural does not exist."
—Camille Flammarion*

"The figures are shocking. Three quarters of the American population literally believe in religious miracles. The numbers who believe in the devil, in resurrection, in God doing this and that—it's astonishing. These numbers aren't duplicated anywhere else in the industrial world. You'd have to maybe go to mosques in Iran or do a poll among old ladies in Sicily to get numbers like this. Yet this is the American population."
—Noam Chomsky.

"There is more religion in man's science than there is science in his religion."
—Henry David Thoreau

"I am so far beyond atheism, there isn't a word in the English language dictionary to describe me."
—Harlan Ellison

"Who says I am not under the special protection of God?"
—Adolf Hitler

---

* Names in SMALL CAPS indicate figures that are quoted in the main section of the book.

THE WORLD (NOT JUST AMERICA) is deeply divided. The main fault line is where the tectonic plates of religion and of reason/secularism/modernity/science/Enlightenment meet and grind against each other, making an absolutely *unbearable* noise. It's sort of like . . . forget it, I can't describe it.

My aim in compiling *The Quotable Atheist* was to heal our broken planet, essentially by eliminating the religious part. Not with nuclear weapons or lesser acts of mass murder, no—that's the *religious* style, nowadays, in certain quarters—but through argument, persuasion, and most of all (since I know perfectly well that argument is *utterly* useless against dumb, blind faith, and just wanted to pay it lip service), the steady application of powerfully abrasive *ridicule* which will slowly but surely *erode* away the offending continent.

I'm serious. Do I really believe this book will convert believers and turn them from the path of self-righteousness to the path of righteousness? Yes. A few. Three, I estimate. Two for sure. But the point is this:

For years, millions of fine, upstanding American atheists and agnostics have watched and stewed as the religious right expanded its influence throughout public life, and as America closed its mind and opened its heart to angels, aliens, ghosts, psychics, Jesus, astrology, Kabbalah, Genesis, Revelation. . . . As SAM HARRIS wrote in *The End of Faith*: "Unreason is now ascendant in the United States—in our schools, in our courts, and in each branch of the federal government. Only 28 percent of Americans believe in evolution; 68 percent believe in Satan. Ignorance in this degree, concentrated in both the head and belly of a lumbering superpower, is now a problem for the entire world."

Meanwhile, religion continues to be granted far too much respect and too little critical examination in our culture and mainstream media. We need to change the cultural climate so as to make supernatural, occult, and faith-based claptrap feel unwelcome and to make adults ashamed of the blithe surrender of their otherwise sound minds to idiocy. We need climate change. Bullshit levels are rising globally, threatening to submerge intellectually low-lying areas. Much of the United States is already inundated. Temperatures are rising; IQs are dropping. Four of the five stupidest years on record have occurred since 2000.

I would of course have preferred a declaration by the president of the United States—purportedly God's messenger on earth—stating that neither God nor WMDs ever existed and that most religious beliefs are untrue and harmful, and urging citizens to bring their minds back up at least to an eighteenth-century stage of development. (I have proposed this plan in a letter to George W. Bush, but haven't heard back yet. They must be hashing out the details.)

Failing that, it is up to atheist/secularist groups and individuals to do what we can to stop global worming (people groveling like worms before nonexistent deities). That's where this book comes in.

As a number of these collected quotes say (far more wittily): Religion in general is based on falsehoods—comforting beliefs in a heavenly parent or big brother; hopes of surviving death—and on utility or expedience: socially cohesive tribal myths; politically useful codes of law and behavior; divine ordination of rulers (including certain presidents); attempts to explain, influence, or placate nature and the elements; the wish to raise ourselves above (i.e., deny our place among) the animals. Religion may help people feel their lives have a loftier purpose than the mere satisfaction of material wants and sensual desires, but it does it with smoke and mirrors, at the cost of our respect for truth and of our integrity and dignity.

Truth has never needed to win converts forcibly. ("Say it—'I believe in the second law of thermodynamics'—*say it,* infidel, or die!") Religion can only demonstrate a miracle; threaten eternal damnation; offer delightful, virginful visions of heaven; promise health, wealth, social or political benefits, or relief from guilt, loneliness, and emptiness; or offer as the other alternative a swift beheading or slow burning.

Persecution and violence have been the rule throughout religious history. Religious tolerance has grown just to the extent that we have become less religious. Religious authority has always sought to obstruct scientific research and education; to control and censor art and literature; to impose rules of behavior that may have made sense centuries or millennia earlier, and probably not even then; and to support rulers and governments, however cruel and oppressive, in exchange for the preservation of its own privileges and wealth. In short, religion has been doing more harm than good for the last, oh, 2,007 years or so. Definitely the last 1,427. Possibly the last 5,767. It has far outlived any beneficial purpose it ever had. (For achieving oneness with the universe, for example, we now have pills.)

Yet religion not only persists; after retreating nicely for a few hundred years, it's back literally with a vengeance—September 11, 2001 being the outstanding recent example. (See quote by Monsignor LORENZO ALBACETE.) In American politics, religion has been rearing its ugly head more rearingly in the past decade or two—and particularly since January 20, 2001—than in the entire history of the United States. And our culture? My God. Make Jesus the center of your movie or novel, make angels, ghosts, witches, vampires, psychics or UFOs the theme of your TV series, and your pecuniary prayers are answered.

If a thinking person of a century ago were told that the next hundred years would see a war in which millions of Jews were murdered out of an originally religious hatred; another war, basically over religion, on European soil (the former Yugoslavia); Middle-Eastern countries still under theocratic rule; enormously popular Islamist groups waging a worldwide jihad; millions of Chinese Falun Gong devotees following a self-anointed savior who also claims the abilities to levitate and to become invisible; arena-sized churches springing up all over the United States; as few as 28 percent of Americans believing evolution is a fact, and 13 percent or fewer believing it occurred through natural selection, unguided by God; the U.S. government dominated by professed evangelical or born-again Christians; Christian fundamentalists holding effective veto power over Supreme Court nominations; and the Oval Office occupied by a man who has affirmed the impossibility of a non-Christian entering heaven[*]—that thinking person might well feel that all the

---

[*] Speaking of the devil, recent U.S. politics illustrates how the habit of faith spills over into politics, fostering blind faith in the politician who advertises himself as a man of God, regardless of his ignorance, incompetence, moral corruption, and/or criminality.

intellectual progress of the previous three or four centuries had been for nought.

We must also count false belief as in itself an irreducibly bad thing, and regard respect for truth supported by reason and evidence as a fundamental human obligation. We should view willful ignorance and stupidity as—I almost said a "sin"; call it an offense against life. And to my mind, to fail to learn about and feel *reverence* for what science has discovered—to refuse to see in the 100-trillion-mile, 100-billion-galaxy extent of the observable universe, the strange-beyond-comprehension subatomic universe, and the possibility of *other* universes (separate God for each or same one for all?)—to refuse to recognize *there* the proper objects of our *religious* attention, if you will, and instead continue to hold barbaric tribal myths as sacred—I call *that* willful ignorance and stupidity. I call it unpardonable *ingratitude* toward the generations of scientists over the centuries who have labored to discover, bit by painstaking bit, real, verifiable knowledge about the world we live in; actual, precious *facts* that have not only made our lives longer, healthier, safer, and more pleasant but are in and of themselves worthy of profound respect, if not, as I said, reverence.

As for belief in any form of creationism: That insults nature itself by denying or ignoring its ability to structure, organize, and build itself. (See LEE SMOLIN.) The evolution of chemical stuff into what we call life (and by the way, it's time we recognized there is no real duality of living and nonliving, let alone between human and animal life) is almost certainly an *inevitable* development from the properties— the geometry, physics, and statistics—of nature's most fundamental stuff. Science presents us at every turn, why, on every boring text-book page, with a fantastically strange world—the 99.99999 percent of our own world that we can't or don't perceive in everyday life. Every significant discovery about nature is, in its sudden, stark reality and radical other-ness, rather like an encounter with an extraterrestrial. (If I had a nickel for every one of *those* I've had.) It's an insult to both science and nature to take no interest in the "miracle" of *reality* while lapping up Tales of the Supernatural, Testaments 1 and 2. Biblical religion adds injury to these insults by teaching that the world was created for humans' exclusive benefit and exploitation, thus sanctioning environmental wantonness and vandalism.

Complaints about the collateral damage caused by religion might almost be beside the point if the beliefs in question—in particular, the belief in God—were true. The thing is—and this is a very important point—there *is* no God. (By writing that, I've just broken Massachussetts law as well as British law. I hope I'm proud of myself.) I mean there exists nothing similar enough to any of the traditional ideas of God to warrant calling it "God." If one means something very different, one should call it something else—*and* explain why we should *worship* it. The pantheist or Spinozist statement "God is nature" or "nature is God," for example—if it merely means, "let's call nature 'God'"—is pointless; to the extent that it suggests that nature has some sort of conscious, purposeful personality, it is false and misleading. Nature certainly never promised Abraham the land of Canaan, spoke to Moses atop Mount Sinai, begat a child by a virgin, dictated a message to the Prophet Muhammad, or revealed to me my divine nature and mission. *God* did those things.

What I really mean, of course, is that there is no reason to believe that there exists a superintelligent, purposeful, conscious—in short, liberal-like—being that created and/or rules the universe; and I have come to earth to tell you that no reason to believe is good reason not only *not to* believe but to *dis*believe.

It's time, I suppose, for the tedious business of defining terms. Feel free to skip a paragraph ahead. *Vaya con dios.*

I'm using "atheist" as shorthand to indicate any nonbeliever in the existence of a deity or deities. This includes both the agnostic, who thinks there's not enough evidence to decide one way or the other—and therefore remains a *nonbeliever;* the atheist proper who goes further and *believes* the Supreme Being *does not* exist—i.e., *dis*believes; and the nontheist, who considers the question meaningless and irrelevant (whereas an agnostic may consider the question unanswered but very important). Throughout the eighteenth century, "Spinozist" or pantheist was regarded as equivalent to "atheist"; I'm using "atheist" in this broad sense. I could have chosen as my catchall term "nontheist" or "agnostic" or the recently coined "Bright," which includes all of the above plus skeptics, "ignostics" (see SHERWIN WINE) and, for all I know, Flying Spaghetti Monsterists (see BOBBY HENDERSON). I just think "atheist" has more sales appeal, in more than one sense.

Much bewilderment arises simply because we ask illogical or meaningless questions—questions that are merely figments of human behavior, perceptions, or language (as philosophers from KANT TO WITTGENSTEIN taught). I suspect, for example, that the "why" in that ultimate question "Why does anything exist?" has no meaning outside of human language, and that the question is no different in type than "Why does $5 = 5$?" Yet people ask it, and the most popular answer throughout history has of course been "God." (Oh, the trouble words cause us! Get rid of them *all*, I say. Imagine there's no language. . . . It's easy if you try. . . . No grammar below us. . . . Above us, no "why.". . . You may say I'm a dreamer. . . .) But to jump to the conclusion—to arrogantly presume to know—that an intelligent, conscious, and purposeful Creator is the answer to that supreme Why makes infinitely less sense than to simply and humbly continue to gain what actual knowledge we can—and meanwhile to acknowledge that there is so far no sign whatsoever of Godot.

"You're missing the point. Faith isn't about numbers or logic. It's about what we know and feel *inside*." That of course is precisely the problem. It means you can believe just about anything. God *did* clearly promise all the land between the Mediterranean and the Euphrates to the children of Israel. Jesus *was*—or Reverend Moon *is*—the incarnation of God on Earth. Psychiatry *does* kill. Infidels must *be* killed. Say what you will about atheism—no one has ever slaughtered in its name. (Stalin's principles were not so lofty. As for HITLER—look him up in this book.)

Yet I confess, I've got a soft spot for "spirituality." Why those quote marks? The trouble with "spiritual" is that, at least according to the diagram in my manual, there aren't two kinds or levels of reality—a physical world and a second, separate, spiritual one. "The supernatural does not exist." Or, if you prefer (and I rather do), it's *all* "supernatural." The very fact of existence *feels* supernatural. At least it cries out for explanation. We are here, apparently, and we don't know why. You can't just call that "natural" and go on your merry way.

I even have a couple of nice things to say about organized religion. I respect, indeed feel awed and somewhat shamed by, the seriousness, depth, and commitment of the genuinely religious life. At its best, it ensures at least a weekly reminder (in church, synagogue, mosque, or hogan) of the Larger Scheme of Things. It can hold families and communities together. Strong, vibrant communities like Jonestown and Heaven's Gate.* Its observances and holy days help to structure, give meaning to, and, from the believer's point of view, sanctify life. It *can* make the adherent a better, kinder, gentler person.

But to qualify for these benefits, which may also include health insurance, you're supposed to *believe* all kinds of bizarre, crazy stuff— beginning, of course, with the Big Guy upstairs. And apart from the iffy benefits, these beliefs bring heavy costs. "Nearly all religions have traditions of fighting for one's faith and supernatural rewards for martyrdom," wrote VANESSA BAIRD. "You have to believe it to do it. . . . So religion does not simply justify violence—the suicide bomber's God, like that of George [W.] Bush, orders it."

But *do* you have to believe? I think most religionists are cheating. Many readily admit that they believe at most in some nebulous Higher Power or Spirit (the Force is also popular these days—see BBC ONLINE) and that for them, religion is mainly about tradition, family, feeling all warm and fuzzy—or perhaps about getting sober. But I contend that even fundamentalist Christians and Muslims by and large don't *really* believe in God. Not the way a European of the Middle Ages could believe—before everyone knew the earth isn't the center of the universe, that microbes, not sin, cause disease, that humans are descended from apelike animals (does anyone today *really* not believe that?), and that thunder . . . what does cause thunder, anyway?—in short, the way a Joan of Arc could believe. As philosopher DANIEL DENNETT puts it (almost echoing Bishop JOHN SPALDING): "Most people in the West who say they believe in God actually *believe in belief* in God." They like the *idea*.

---

* See JOHN SHELBY SPONG, final quote.

Faith today is a flag. The current worldwide religious revival is a rebellion against modernity, in the name of identity, tribe, and tradition—a protest against the paving over of local and national traditions by the steamroller of global commercial-consumer culture. (Ground Zero was after all the *World Trade* Center.) Religion has rushed in to fill the genuine spiritual vacuum created by modernity and capitalism—the archenemy of *all* tradition, including religion, as conservatives once understood. Osama bin Laden is the bastard offspring of LOCKE, Adam Smith, BENTHAM, MILL, VOLTAIRE, BONAPARTE. . . .

The problem, however, isn't that science, secularism, and Enlightenment ideas have become the modern religion, but rather that they haven't. While fundamentally antagonistic toward science, the current antimodernist, *post*modernist religious reaction isn't against a rival belief system or philosophy, so much as against an *absence* of strongly held values—a spiritual-philosophical void that business seeks to fill with toys, amusements, and celebrity idolatry, replacing old-fashioned ideals, with a cool cynicism, irony, and disdain for any kind of seriousness and passion—religious, political, emotional, or intellectual. TV and advertising have bred these virtues out of our culture because they were rivals to the "values" of consumption; after all, if people start feeling their lives are rich and meaningful enough without having to *buy* stuff, *then* where will we be?

Ironically, atheism in its best sense—as a positive, vigorous philosophy of life rather than just a rejection of—confronts the same unseriousness. Your average nonbeliever—and I know him well—just can't be *bothered* thinking through what he believes and doesn't believe. He isn't *against* religion so much as uninterested. To him, the *Origin-of-Species*-thumping atheist orators of yester-century are as antique as an old-time revival meetin'. At the same time, multiculturalism has taught us that while it may be okay to trash fundamentalist Christianity (a freedom that this book will eagerly avail itself of), we mustn't presume to tell non-Western societies what *they* should and shouldn't believe.

But with creeping theocracy at home and religious fanaticism on the loose everywhere, this is no time for ironic detachment (read, intellectual laziness). It's time to get post-postmodern about religion: less indifferent, less respectful, less relativistic—frankly, less tolerant—toward religion *per se*, not just its more "extreme" (read *pure*) manifestions. The problem with religious moderation is that it helps keep barbaric, scripture-based beliefs respectable. Religious moderation, Sam Harris notes, "is the product of *secular* knowledge and scriptural ignorance." Moderates ignore or pay lip service to scriptural injunctions; Osama bin Laden acts on them. In an age of suitcase bombs, tribalistic irrationalism in any degree is no longer just a harmless anachronism.

Atheists, for our part, can be as closed-minded, humorless, self-righteous, and *evangelical* as religionists. Throughout the ages, nonbelief has served some atheists as a means of feeling superior to the credulous crowd, the moronic, manipulable, mesmerized masses. (Not that the feeling isn't justified.) To some atheists, one must be a strict materialist, free of any taint of spirituality, if one is to remain above suspicion of un-atheistic activities. At least one atheist Web site displays an amusingly McCarthyite zeal for rooting out religious or spiritual sympathies among supposed nonbelievers.

But spiritual appetites are real (see HUBERT HARRISON, last quote), worthy of respect, and of course a begging to be milked for profits. What's needed is more than just the discrediting and utter humiliation of religion as we know it, but a counter-religion, if you will—a positive, assertive, science-based secularism that amounts to, and isn't squeamish about being, an alternate way forward "spiritually." Science, after all, arose out of mysticism and sorcery, with which it shares the thirst for higher knowledge and for mastery over nature. In fact, in many ways science has been growing more mystical again. I believe that, just as certain eighteenth- and nineteenth-century Spinozists envisioned, science is destined to replace, or if you prefer, remerge with, religion. It may need to be called something else. (The First Church of Something Else? Has a nice, tax-exempt ring to it.)

Actually, there's nothing wrong even with the word "religion." Put a new engine in it, clean it up, get rid of that *smell,* and it'll be good as new. All the word means, after all, is "re-tying"—*reconnecting* with our cosmic origins, with the LST (Larger Scheme of Things)—and nothing's wrong with that. LST will open your *mind,* man.

Accordingly, this book is intended not only to inspire, amuse, and pander to nonbelievers but also to challenge them—to be thought provoking to atheists and theists alike. Hence, I didn't trouble much about ideological purity, and included agnostics, ignostics, eggnostics—even theists and theocrats. (In fish-nor-fowl cases such as America's deist Founding Fathers, I of course carefully cherry-picked the quotes that put them most decisively in the antireligious camp.) I favored quotes that bespeak a really open mind—open even to the appalling possibility of a consciousness and purpose in nature. (My garden, God knows, seems to have a mind of its own.) I strove to include figures from all periods and walks of life, especially if they added celebrity and sex appeal. But above all, in sifting through thousands of quotes in books and on the Internet, I looked for the most brilliant, penetrating, and funny. The words of ANDRÉ MALRAUX, "Be careful—with quotations you can damn anything," were my constant inspiration.

Atheism *should* mean getting, and laughing at, the mother of all jokes: that we're animals—animated mud—agglomerations of molecules that have evolved the ability to reproduce themselves and have formed into large colonies that *we* regard as plants, animals, people (but which our selfish genes "regard" as cities and vehicles)— colonies whose Weblike web of neural interconnections can even produce an impression of consciousness. We, the "most evolved" of these colonies, anointed ourselves the center and purpose of the universe, only to discover that we're not, but that the likes of us have apparently been left in charge of a planet in a universe *without* purpose, center, leader, or moral authority; that we're all, as the Firesign Theater once put it, "just bozos on this bus." Which suggests we'd better put aside tribal myths, absolutist truths, and obsessional mass neuroses and learn to get along.

SECULAR * INFIDEL * NONBELIEVER * HUMANIST * RATIONALIST * FREETHINKER * AGNOSTIC * GODLESS * HERETIC * ATHEIST *

A

David Aaronovitch (1954–), *British journalist, broadcaster and author. Former communist. Twice winner of the* GEORGE ORWELL *prize for political journalism. For making what he called "a left-wing case for supporting the overthrow of a fascist regime [Saddam Hussein's] . . . I have had the almost astral experience of finding myself excommunicated" and labeled a neoconservative by the left—while continuing to be criticized by the right as a liberal. Also accused of Islamophobia for criticizing Muslim organizations for antigay, antifeminist, and illiberal positions. Wrote of his embarrassment at being in the same room as his young daughter when the TV news reported that President Clinton had received oral sex in an Oval Office vestibule—until she asked, "Daddy, what's a vestibule?"*

> "For people with God on their side, monotheists are a touchy lot. . . . In Exodus, Moses gets the tribe of Levi to go with 'sword at side' and massacre 3,000 calf-worshippers. And we are supposed to celebrate such a violation of the freedom to worship? . . . Why are they so touchy? The problem is partly that all monotheisms are, by their nature, anti-pluralistic. They've got the one true God, and the very latest valid version of his thoughts. It is asking a lot of monotheisms to coexist with other faiths and views. Paganism, on the other hand, is much better suited to modern ideas of tolerance and human rights. Under polytheism you can choose your own god overtly."

Edward Abbey (1927–1989), *American author/environmental advocate. Wrote most memorably about the Western deserts where he was once a park ranger. His novel* The Monkey Wrench Gang, *about a group of "eco-warriors" who sabotage development projects, is said to have inspired the formation of radical environmental groups like Earth First. Writer Larry McMurtry called him "the THOREAU of the American West"; sometimes called "the desert anarchist." Specified that he wanted his body to fertilize "a cactus, a cliffrose, a sagebrush or a tree."*

"Whatever we cannot easily understand we call God; this saves much wear and tear on the brain tissues. . . . Belief in the supernatural reflects a failure of the imagination."

"Fantastic doctrines (like Christianity or Islam or Marxism) require unanimity of belief. One dissenter casts doubt on the creed of millions. Thus the fear and the hate; thus the torture chamber, the iron stake, the gallows, the labor camp, the psychiatric ward."

Clark Adams, *Public relations director for Internet Infidels, creators of the Secular Web; cofounder of the Secular Coalition for America; moderator of the alt.atheism newsgroup; organizer of the annual Lollapalooza of Freethought at the Freedom from Religion Foundation's Lake Hypatia*resort in Alabama, "right in the buckle of the Bible Belt."*

"If atheism is a religion, then health is a disease."

Douglas Adams (1952–2001), *British radio dramatist; author of the* Hitchhiker's Guide to the Galaxy *series. Said the idea came to him while he lay drunk in a field in Austria, gazing at the stars. He was carrying a book called* The Hitchhiker's Guide to Europe. *Previous occupations included chicken-shed cleaner, bodyguard for an Arab royal family, and guitarist for Pink Floyd. Was six feet tall by age 12. A giant among freethinkers. Professed "radical atheist."*

---

* See HYPATIA BRADLAUGH BONNER

"There is a theory that states that if ever anybody discovers exactly what the universe is for and why it is here, it will instantly disappear and be replaced by something even more bizarre and inexplicable . . . There is another theory that states that this has already happened."

John Adams (1735–1826), *Founding Father and second U.S. president. A Deist, like many of the F.F.'s—including* THOMAS JEFFERSON *and* GEORGE WASHINGTON. *Deists rejected organized religion and the divinity of Christ and held that reason is the path to knowledge, including knowledge of God. Some saw God as a clockmaker who created the world but does not intervene in it, or does so only as a subtle force. Others believed God is the universe. In 1831 an Episcopal minister complained: "Among all our presidents from Washington downward, not one was a professor of religion, at least not of more than Unitarianism."*

"God has infinite wisdom, goodness and power; he created the universe. . . . He created this speck of dirt and the human species for his glory; and with deliberate design of making nine-tenths of our species miserable for ever for his glory. This is the doctrine of Christian theologians, in general, ten to one. . . . Wretch! What is his glory? Is he ambitious? Does he want promotion? Is he vain, tickled with adulation, exulting and triumphing in his power and the sweetness of his vengeance? Pardon me, my Maker, for these awful questions."

"Twenty times in the course of my late reading, have I been upon the point of breaking out: This would be the best of all possible worlds, if there were no religion in it!"

"Even since the Reformation, when or where has existed a Protestant or dissenting sect who would tolerate *a free inquiry?*"

From the Treaty of Tripoli, *ratified unanimously in the Senate and signed by Adams into law, 1797:* "The United States of America is in no sense founded on the Christian Religion."

"Who does not see that the same authority which can establish Christianity, in exclusion of all other Religions, may establish with the same ease any particular sect of Christians, in exclusion of all other Sects?"

"In the formation of the American government . . . it will never be pretended that any persons employed in that service had interviews with the gods, or were in any degree under the influence of heaven."

Never say never:

"I trust God speaks through me."—*George W. Bush.*

Scott Adams (1957–), *American cartoonist-satirist. Creator of the* Dilbert *comic strip; CEO of Scott Adams Foods, Inc., makers of the Dilberito, America's favorite microwavable vegetarian burrito.*

"Nothing defines humans better than their willingness to do irrational things in the pursuit of phenomenally unlikely payoffs. This is the principle of lotteries, dating, and religion."

Wayne Adkins (1948–), *American military officer. Served in Iraq in 2004–2005. Former fundamentalist Christian who became an atheist after studying to become a Baptist preacher. "Instead, the more I studied the Bible, the more problems I uncovered and eventually . . . I stopped believing." Maintains the Naked Emperor atheist Web site.*

"How do you choose between believing in Jesus, Bigfoot, leprechauns, witchcraft, Islam, alien abductions, the Tooth Fairy, gold at the end of the rainbow or the myriad other assertions that people have made over the course of human history? [Faith is] like rolling the dice and hoping you have placed your faith in a true proposition. . . . However, if you are still inclined to place faith in an un-provable assertion, I am God, send me money."

"Adonis" (born Ali Ahmad Sa'id, 1930–), *Syrian-born Lebanese-Arab-French poet/editor/publisher. Did not attend school, see a car, or listen to a radio until age 12. His father, a farmer and imam, gave him a traditional Islamic education. Later attended a French lycée and studied law and philosophy at Syrian University. Often mentioned as a Nobel candidate.*

"The religious interpretations that compel Muslim women to wear the veil in secular countries where church and state have long been separated and where equality of the sexes is firmly established, reveals a mentality that is not content merely with veiling woman, but seeks to shroud man, society, life in general—to pull the veil over the eyes of reason itself."

Decca Aitkenhead (1971–), *British journalist and broadcaster. On "an almighty row" that broke out about the teaching of creationism in a British school in 2002:*

"'Rational' Christians fell over themselves to make it plain that they were much too sensible to believe such fairy tales . . . about God creating the world in six days. What a preposterous suggestion! Where was the science in that? Everyone knew the story of Genesis was just a rhetorical flourish. God created evolution. Now, not a month later, the same Christians ask us to believe the story of Easter. . . . Attributing God's authorship to either version of ["creation"] events comes down to the same thing: you believe in a supernatural power. . . . Trying to defend religion by invoking science is like claiming that three plus four equals ice cream."

Monsignor Lorenzo Albacete, *professor of theology, St. Joseph's Seminary, New York.*

"From the first moment I looked into that horror on September 11th, into that fireball, into that explosion of horror, I knew it, I recognized an old companion. I recognized religion."

Ayaan Hirsi Ali (1969–), *Somali-Muslim-born Dutch author, filmmaker, human rights and women's rights activist, and former member of the Dutch parliament (2003–2006). Has received repeated death threats from Islamists for her public rejection of Islam and its subjugation of women and for her coproduction of a film on the latter subject with Theo Van Gogh, a Dutch filmmaker (and great-great-nephew of VINCENT VAN GOGH) who was later murdered by an Islamist extremist/film critic. A letter containing a*

death threat to Hirsi Ali, along with a rant about Jewish conspiracies and a vow that America, Europe, and the Netherlands "will go down," was pinned to Van Gogh's body with a knife. In her youth Hirsi Ali wore a hijab (full head-scarf) and supported the Islamist Muslim Brotherhood. Fleeing a forced marriage, she received political asylum in the Netherlands. Inspired by the Atheist Manifesto *of Dutch philosopher Herman Philipse, she renounced Islam and became an atheist. Author of* The Caged Virgin: An Emancipation Proclamation for Women and Islam. *Named one of the 100 Most Influential Persons of the World by* Time *magazine in 2005. Nominated for the 2006 Nobel Peace Prize. Can denounce religious beastliness in fluent English, Somali, Arabic, Swahili, Amharic, and Dutch. On first hearing, as a schoolgirl, of* SALMAN'S RUSHDIE'S *Satanic Verses:*

> "We had heard that there was this book, and that the author had said something horrible about the Prophet, which was extremely blasphemous. And the first thought that came into my head was simply, 'Oh, he must be killed.'"

On the Great Dane-ish Cartoon Controversy of 2005:

> "I do not seek to offend religious sentiment, but I will not submit to tyranny. Demanding that people who do not accept Muhammad's teachings should refrain from drawing him is not a request for respect but a demand for submission."

Tariq Ali (1943–), *Pakistani-born British historian, journalist, film-maker, antiwar activist, atheist. Grew up in a communist family in Lahore.*

> "From the age of five or six I was an agnostic. At twelve I became a staunch atheist. . . . But [Muslim culture] has enriched my life. . . . The historian Isaac Deutscher used to refer to himself as a non-Jewish Jew, identifying himself with a long tradition of intellectual scepticism, symbolised by SPINOZA, FREUD and MARX. I have . . . on occasion, described myself as a non-Muslim Muslim."

From his book The Clash of Fundamentalisms:

> "I want to write of the setting, of the history that preceded [9/11], of . . . an increasingly parochial culture that celebrates the virtues of ignorance [and] promotes a cult of stupidity . . . a

world in which escapist fantasies of every sort are encouraged from above."

Amy Alkon (c. 1970–), *American "Advice Goddess." In her column of that name, syndicated in over 100 papers, she enjoys disparaging astrologers, psychics, and other serious, dedicated scientists. From a 2003 column titled "We're The Chosen People And You Suck!":*

> "That anybody believes in God—just because somebody severe-looking told them God exists—is really a hoot. . . . Some guy who looks like Charlton Heston clutching the Ten Commandments is passing judgment on the universe. . . . Like the guy would even have time to care about your pathetic little life: 'Hmm, Amy stiffed the coffee bar on a tip today. I suppose I'll have to smite her after I finish my breakfast.'. . . If people really just wanted a framework to be good, all they'd need to do is live by the tenets of the religion I've created (and feel free to call it Amyism): 1. Be kind. 2. Be ethical. 3. Be rational. 4. Live as if a piano could fall on your head at any moment . . . 5. Leave the campground better than you found it."

Ethan Allen (1738–1789), *American Revolutionary War hero. After publishing an attack on Calvinist Christianity titled* Reason, the Only Oracle of Man, *a minister called him "Antichrist." When asked at his wedding to pledge to live with his bride "agreeable to the laws of God," he halted the ceremony in protest. Attempted during the war to capture Montreal, my hometown, but I forgive him his trespasses.*

> "In those parts of the world where learning and science have prevailed, miracles have ceased; but in those parts of it as are barbarous and ignorant, miracles are still in vogue."

Steve Allen (1921–2000), *American comedian, musician, songwriter, screenwriter, creator and original host of NBC's* Tonight Show, *and author of 43 books, including three on his rejection of biblical religion. Raised*

Catholic. As an entertainer who often stayed in hotels, he discovered his abhorrence of the Old Testament God while reading the Gideon Bibles. So you see, they do do good.

"Both the existence and the nonexistence of God seem in some respects preposterous. I accept the probability that there is some kind of divine force, however, because that appears to me the least preposterous assumption of the two."

"We hear much, in recent years, of a return to religion. . . . There is also an especially disturbing proliferation of bizarre cults and freako churches . . . whose belief systems are intellectually on a par with the mindset of supermarket tabloids. And even within religious groups that are respectful of scientific evidence, not to mention common sense, it is the irrational and superstitious wings that are flourishing. . . . Among the hundreds of millions of believers in Islam, it is the most violent and fanatical elements that are flourishing." *(Written in 1994.)*

"Few if any articulate atheists, agnostics, or secular humanists have been attracted to cults." *Allen's son joined a Jesus cult in the 1970s. (May have been called "Christianity"—not sure.)*

"I believe it is the imposition of a dictatorship that increasing numbers on the Christian Right now wish to construct in the United States. . . . Fundamentalist Christians [who] believe that the Bible is reliable as history and science are no longer content with teaching their freely-gathered congregations. . . . When they insist on having historical and scientific errors taught in our nation's public schools, then they must be opposed by all legal means." *And if those fail?*

Woody Allen *(born Allen Stewart Konigsberg, 1935–), comedian, film director, screenwriter, actor. First published joke: "I am two with Nature." He's definitely also two with God. "The Jewish voice in mainstream society is often cynical, skeptical and philosophically materialist."*[2]* (Also see LENNIE BRUCE, JACKIE MASON, HENNY YOUNGMAN.)*

"Not only is God dead, but just try to find a plumber on weekends."

* See Sources, p. 335

"There's no way to prove that there is no God. You just have to take it on faith."

"I do not believe in an afterlife, although I am bringing a change of underwear."

"I don't want to achieve immortality through my work . . . I want to achieve immortality by not dying."

Thomas Altizer (1927–), *American theologian. A leader of the 1960s "death of God" theological movement (a.k.a. theothanatology). Author of* The Gospel of Christian Atheism *and* The Death of God. *God packed up and moved to Earth as Christ, and although the latter died, the former's spirit has been living* here—*and not "out there"—ever since. Or something like that.*

"Only by accepting and even willing the death of God in our experience can we be liberated from a transcendent beyond, an alien beyond which has been emptied and darkened by God's self-alienation in Christ." (Also see RICHARD RUBENSTEIN.)

Henri Frédéric Amiel (1821–1881), *Swiss poet and philosopher.*

"We are always making God our accomplice so that we may legalise our own inequities. Every successful massacre is consecrated by a Te Deum, and the clergy have never been wanting in benedictions for any victorious enormity."

"The efficacy of religion lies precisely in what is not rational. . . . Religion attracts more devotion according as it demands more faith—that is to say, as it becomes more incredible to the profane mind."

"A belief is not true because it is useful."

Martin Amis (1949–), *British novelist. His father, the much better and funnier novelist but reactionary shit Kingsley Amis, criticized Martin's work for "breaking the rules, buggering about with the reader, drawing attention to himself. . . ."*

"Since it is no longer permissible to disparage any single faith or creed, let us start disparaging all of them. To be clear: an ideology is a belief system with an inadequate basis in reality; a religion is a belief system with no basis in reality whatever. Religious belief is without reason and without dignity, and its record is near-universally dreadful. It is straightforward—and never mind, for now, about plagues and famines: if God existed, and if He cared for humankind, He would never have given us religion."

Anaxagoras (c. 500–428 B.C.E.), *Greek philosopher. Regarded the conventional gods "as mythic abstractions endowed with anthropomorphic attributes. His writings led him to a dungeon, charged with impiety."*[1] *Only Pericles' intervention saved him from a death sentence. He had to pay a fine, was banished, and lived his final years in exile.*

"Everything has a natural explanation. The moon is not a god but a great rock and the sun a hot rock."

Anaximander (c. 610–546 B.C.E.), *Greek philosopher, journalist, and media personality. Was named Proto-Darwinian of the Year in 561 B.C.E.*

"Living creatures arose from the moist element as it was evaporated by the sun. Man was like another animal, namely a fish, in the beginning."

Peter A. Angeles, *American philosophy professor. Author/editor of* Critiques of God: Making the Case against God. *Writer and host of a radio show,* The Children's Story Time, *in which he no doubt corrupts small children with his atheist filth.*

"What was God doing . . . for an eternity . . . before He created the universe ex nihilo? God existed by Himself through an eternity . . . without needing a universe. Why did He suddenly desire to create the universe?" *(He got hungry and needed someplace to order pizza from?)*

Natalie Angier (1958–), *Pulitzer-winning* New York Times *science writer. From her article "Confessions Of A Lonely Atheist,"* New York Times Magazine, *2001:*

> "[Today,] nothing seems as despised, illicit and un-American as atheism. . . . So, I'll out myself. I'm an Atheist. I don't believe in God, Gods, Godlets or any sort of higher power beyond the universe itself, which seems quite high and powerful enough to me. I don't believe in life after death, channeled chat rooms with the dead, reincarnation, telekinesis or any miracles but the miracle of life and consciousness, which again strike me as miracles in nearly obscene abundance. . . . I'm convinced that the world as we see it was shaped by the again genuinely miraculous, let's even say transcendent, hand of evolution through natural selection."

Anonymous (536 B.C.E.–2006 C.E.), *prolific author of T-shirt and bumper sticker slogans.*

> "Philosophy is questions that may never be answered. Religion is answers that may never be questioned."

> "Morality is doing what is right no matter what you are told. Religion is doing what you are told no matter what is right."

> "Education and religion are two things not regulated by supply and demand. The less of either the people have, the less they want."—Charlotte Observer, *1897*

> "The mind of the fundamentalist is like the pupil of the eye: the more light you pour on it, the more it will contract."

> "Christian Fundamentalism: The doctrine that there is an absolutely powerful, infinitely knowledgeable, universe spanning entity that is deeply and personally concerned about my sex life."

> "If God doesn't like the way I live, let him tell me, not you."

> "Out of convicted rapists, 57 percent admitted to reading pornography; 95 percent admitted to reading the Bible."

> "Blasphemy is a victimless crime."

"Give a man a fish and you'll feed him for a day. Give him a religion and he'll starve to death while praying for a fish."

"Only sheep need a shepherd."

"Why be born again, when you can just grow up?"

"Christian: 'I'll pray for you.' Atheist: 'Then I'll think for both of us.'"

Nineteenth-century Orthodox Jews' reproach to Enlightenment and Reform Jews: "The Torah teaches, 'Hear, O Israel' not 'Think, O Israel.'"

"Power corrupts. Absolute power corrupts absolutely. God is all-powerful."

"If forgiveness is divine, why is there a hell?"

"If you see a blind man, run up and kick him. Why should you be kinder than God?"—*Old Iranian proverb*

"Nothing that would invent a mosquito is worthy of anything but hate."

"Organized religion is like organized crime; it preys on peoples' weaknesses, generates huge profits for its operators, and is almost impossible to eradicate."

Ancient Spartan whose confession a Christian priest wanted to take: "Is it to you or to God I am to confess?" "To God." "In that case, man, begone!"

"The only worse liar than a faith healer is his patient."

"The 'religious right' aren't and 'scientific creationism' isn't."

"Most people hate the idea of evolution because they realize that if it were working properly, they'd be dead."

Jean Anouilh (1910–1987), *French playwright.*

"Every man thinks God is on his side. The rich and powerful know he is."

Susan B. Anthony (1820–1906), *American women's suffrage and antislavery crusader. Arrested, tried, and (although defended by* MATILDA

JOSLYN GAGE) *found guilty for casting a vote in the 1872 presidential election. Campaigned against abortion—then seen as an imposition forced on women by men. Expelled from the National Labor Union for encouraging women to enter the printing trades while male workers were on strike.*

"I was born a heretic." *(A Quaker, actually.)*

"I distrust those people who know so well what God wants them to do because I notice it always coincides with their own desires."

Brian Appleyard (1951–), *British author.*

"Modernism may be seen as an attempt to reconstruct the world in the absence of God."

Louis Aragon (1897–1982), *French novelist, poet, essayist. Dada and Surrealist movement leader. After the death of his wife in 1970, he revealed his bisexuality and appeared at Paris gay pride parades in a pink convertible.*

"Of all possible sexual perversions, religion is the only one to have ever been scientifically systematized."

Aristophanes (c. 448–385 B.C.E.), *Greek comic dramatist. Prosecuted for libel at least twice. His ungovernable characters say things like:*

"Shrines! Shrines! Surely you don't believe in the gods. What's your argument? Where's your proof?"

"What sort of god is Zeus? Why spout such rubbish? There's no such being as Zeus. . . . Just tell me—where have you ever seen the rain come down without the Clouds being there? If Zeus brings rain, then he should do so when the sky is clear, when there are no Clouds in view." *(But then how would Zeus hide from mortals the fatal sight of his terrible countenance, dummy?)*

Aristotle (c. 384–322 B.C.E.), *Greek philosopher. Did not, as per legend, throw himself into the sea because he couldn't explain the tides;*

to propose giving that "alternative theory" equal time in the classroom is ludicrous.

"Men create gods after their own image, not only with regard to their form but with regard to their mode of life." *(Greek gods had* lifestyles, *which is more than can be said for Jehovah.)*

"A tyrant must put on the appearance of uncommon devotion to religion. Subjects are less apprehensive of illegal treatment from a ruler whom they consider god-fearing and pious [and] less easily move against him."

Karen Armstrong *(1944–), British religious historian and former nun.*

"The statement 'I believe in God' has no objective meaning, as such, but like any other statement only means something in context, when proclaimed by a particular community. . . . A fundamentalist would deny this, since fundamentalism is antihistorical: it believes that Abraham, Moses and the later prophets all experienced their God in exactly the same way as people do today."

"[Fundamentalisms] are embattled forms of spirituality, which have emerged as a response to a perceived crisis. . . . Fundamentalists fear annihilation [by secularism], and try to fortify their beleaguered identity by means of a selective retrieval of certain doctrines and practices of the past."

"The Qur'an reflects the brutal tribal warfare that afflicted Arabia during the early seventh century. . . . The scriptures all bear scars of their violent begetting, so it is easy for extremists to find texts that give a seal of divine approval to hatred."

"The Christian Right today has absorbed the endemic violence in American society: they oppose reform of gun laws, for example, and support the death penalty. They never quote the Sermon on the Mount [*"love thy neighbor . . ."; "judge not, lest ye be judged"*] but base their xenophobia and aggressive theology on Revelation."

Lance Armstrong (1971–), *American cyclist. Seven-time consecutive Tour de France winner (1999–2005) and cancer survivor. Self-described as "middle to left" and "against mixing up State and Church." Writing about the night before he underwent brain surgery:*

> "I asked myself what I believed. I had never prayed a lot. I hoped hard, I wished hard, but I didn't pray. I had developed a certain distrust of organized religion growing up, but I felt I had the capacity to be a spiritual person. . . . I believed I had a responsibility to be a good person . . . fair, honest, hardworking, and honorable. . . . If there was indeed a God at the end of my days, I hoped he didn't say, 'But you were never a Christian, so you're going the other way from heaven.' If so, I was going to reply, 'You know what? You're right. Fine.'"

Referring to the organs where his cancer started:

> "If there was a God, I'd still have both nuts."

Liv Arnesen (1953–), *Norwegian explorer and motivational speaker. First woman to ski solo to the South Pole (1994) and all the way across Antarctica (at age 48, with American Ann Bancroft).*

> "I know many people who believe in God, and I expected to find Him on my way to the South Pole if He exists. My religious experiences were very different however, involving [only] myself, nature and the universe." *(Has also explored the Arctic. No Santa.)*

Matthew Arnold (1822–1888), *English poet and foremost literary critic of his era. According to T. S. Eliot, Arnold wanted "to get all the emotional kick out of Christianity one can, without the bother of believing it. . . . The total effect of Arnold's philosophy is to set up Culture in the place of religion." (Right . . . is there a problem?)*

> "So deeply unsound is the mass of traditions and imaginations of which popular religion consists, that future times will hardly comprehend its audacity in calling those who abjure it atheists."

"The theological faculty of the University of Paris, the leading medieval university, discussed seriously whether Jesus at his ascension had his clothes on or not. If he had not, did he appear before his apostles naked? If he had, what became of the clothes?" *Incredible. They should have been asking where he* bought *his clothes: at an overpriced store or at the home of Jerusalem's best values in men's wear?*

Isaac Asimov (1920–1992), *Russian-born American biochemist, science and science-fiction author. Wrote or edited over 500 books (and an estimated 90,000 letters and postcards). Served as president of the American Humanist Association from 1985 until his death, when he was succeeded by* KURT VONNEGUT. *The asteroid 5020 Asimov is named in his honor.*

"I am an atheist, out and out. It took me a long time to say it. . . . Somehow I felt it was intellectually unrespectable . . . because it assumed knowledge that one didn't have. Somehow it was better to say one was a humanist or agnostic. I don't have the evidence to prove that God doesn't exist, but I so strongly suspect that he doesn't that I don't want to waste my time."

"Properly read, the Bible is the most potent force for atheism ever conceived."

"Imagine the people who believe such things and who are not ashamed to ignore, totally, all the patient findings of thinking minds through all the centuries since the Bible was written. And it is these ignorant people, the most uneducated, the most unimaginative, the most unthinking among us, who would make themselves the guides and leaders of us all; who would force their feeble and childish beliefs on us; who would invade our schools and libraries and homes. I personally resent it bitterly."

"Creationists make it sound like a 'theory' is something you dreamt up after being drunk all night."

"To rebel against a powerful political, economic, religious, or social establishment is very dangerous and very few people do it, except, perhaps, as part of a mob. To rebel against the 'scientific' establishment, however, is the easiest thing in the world, and anyone can do it and feel enormously brave, without risking as much as a hangnail."

"No vision of God and heaven ever experienced by the most exalted prophet can, in my opinion, match the vision of the universe as seen by Newton or Einstein."

Kemal Atatürk (Mustafa Kemal Pasha, 1881–1938), *founder, first president, and pitiless modernizer and secularizer of the Turkish republic. Removed Islam as the state religion; replaced the Arabic alphabet with the Roman, and religious, Arabic-language schooling with secular, Turkish-language schools; established universal suffrage (yes, including women); and, most important, decreed that men abandon the fez in favor of European-style hats; in short, filled Turkey to the brim with modernity. "One of the few positive things Atatürk said about religion was that since his soldiers thought they were going to heaven, they were conveniently willing to die."* [2]

"I have no religion, and at times I wish all religions at the bottom of the sea. He is a weak ruler who needs religion to uphold his government; it is as if he would catch his people in a trap. My people are going to learn the principles of democracy, the dictates of truth, and the teachings of science *[if I have to slaughter every last one of them]*. Superstition must go."

Peter William Atkins (1940–) *British chemist. Author of two of the world's leading chemistry textbooks as well as* Galileo's Finger: The Ten Great Ideas of Science *and* The Creation.

"It is not possible to be intellectually honest and believe in gods. And it is not possible to believe in gods and be a true scientist."

"Religion closes off the central questions of existence by attempting to dissuade us from further enquiry by asserting that we cannot ever hope to comprehend. We are, religion asserts, simply too puny. Through fear of being shown to be vacuous, religion denies the awesome power of human comprehension. It seeks to thwart, by encouraging awe in things unseen, the disclosure of the emptiness of faith. . . . Science opens up the great questions of being to rational discussion. . . . Science, above all, respects the power of the human intellect. . . . Science respects more deeply the potential of humanity than religion ever can."

Rowan "Mr. Bean" Atkinson (1955–), *British actor/comedian. Starred in, and cowrote some of, the richly sacrilegious British TV comedy series* Blackadder. *Stuttered as a child and still has particular trouble with the letter B (as in belief, benediction, bishop). Led a coalition of prominent actors and writers opposed to Britain's Racial and Religious Hatred Bill (which Muslim groups lobbied for) as a threat to freedom of speech and expression:*

> "Having spent a substantial part of my career parodying religious figures from my own Christian background, I am aghast at the notion that it could, in effect, be made illegal to imply ridicule of a religion or to lampoon religious figures. . . . I have always believed that there should be no subject about which one cannot make jokes, religion included. . . . For telling a good and incisive religious joke, you should be praised. For telling a bad one, you should be ridiculed and reviled. The idea that you could be prosecuted for the telling of either is quite fantastic. . . . Comedy takes no prisoners."

A bishop addressing Blackadder (Atkinson):

> "You fiend! Never have I encountered such corrupt and foul-minded perversity. . . . Have you ever considered a career in the church?"

Blackadder's servant, Baldrick, describes his Nativity play woes:

> "At the last moment, the baby playing Jesus died!" "Oh, dear. . . . What did you do?" "Got another Jesus!" "Oh, thank goodness. And his name?" "Spot. . . . There weren't any more children, so we had to settle for a dog instead. . . . Well, it went alright 'til the shepherds came on. See, we hadn't been able to get any real sheep, so we had to stick some wool . . ." " . . . on some other dogs." "Yeah. And the moment Jesus got a wiff of 'em, he's away! . . . So while the angels are singing 'Peace on Earth, goodwill to all men,' Jesus is trying to get one of the sheep to give him a piggyback ride!" "Oh no! . . . Weren't the children upset?" "Nah, they loved it! They want us to do it again next year for Easter. They want to see us nail up the dog."

Sir David Attenborough (1926–), *English broadcaster and naturalist; brother of filmmaker Lord Richard Attenborough. Writer and presenter of nine popular nature documentary TV series. As a BBC Controller,*

commissioned Kenneth Clark's series Civilization *and* JACOB
BRONOWSKI's *Ascent of Man. To be knighted by the queen and have a
species of long-beaked echidna named after you (*Zaglossus attenboroughi*)
yet still not believe in God—I call that ingratitude.*

"I don't know [why we're here]. People sometimes say to me,
'Why don't you admit that the humming bird, the butterfly, the
Bird of Paradise are proof of the wonderful things produced by
Creation?' And I always say, well, when you say that, you've
also got to think of a little boy sitting on a river bank, like here,
in West Africa, that's got a little worm, a living organism, in his
eye and boring through the eyeball and is slowly turning him
blind. The Creator God that you believe in, presumably, also
made that little worm. Now I personally find that difficult to
accommodate. . . ."

Margaret Atwood (1939–), *Canadian novelist, poet, critic. Her
novel* The Handmaid's Tale *(1985) was shortlisted for the Booker Prize
and won the* ARTHUR C. CLARKE *Award for Science Fiction.* The Blind
Assassin *(2000) won the Booker. Named both Canadian and American
Humanist of the year 1987. Invented a device to allow godlike authors to
remotely sign a book while interacting via video and audio.*

Self-description: "A doctrinaire agnostic [which is] different from
someone who doesn't know what they believe. A doctrinaire
agnostic believes quite passionately that there are certain things
that you cannot know, and therefore ought not to make
pronouncements about. In other words, the only things you can
call knowledge are things that can be scientifically tested."

"God is not the voice in the whirlwind. God is the whirlwind."
*(Can we ever be really certain of that? If not, isn't it irresponsible to
make such statements?)*

St. Augustine of Hippo (354–430), *Roman–North African
saint. In his youth, a serious party animal and—to the horror of his mother,
Saint Monica—a follower of the Manichean faith. (He might as well have
brought a Mithraist home.)*

Hope this clears things up: "God always is, nor has He been and is not, nor is but has not been, but as He never will not be; so He never was not."

"Often a non-Christian knows something about the earth, the heavens . . . about the motions and orbits of the stars . . . and this knowledge he holds with certainty from reason and experience. It is thus offensive and disgraceful for an unbeliever to hear a Christian talk nonsense about such things, claiming that what he is saying is based in Scripture." *Tell us about it . . .*

"The good Christian should beware of mathematicians. . . . The danger already exists that mathematicians have made a covenant with the devil to darken the spirit and confine man in the bonds of Hell." *(I hear the mathematicians are in league with the Jews. In fact, see HERMANN BONDI, MAX BORN, JACOB BRONOWSKI.)*

Sir A. J. (Alfred Jules) Ayer (1910–1989), *British philosopher. Author of the classic* Language, Truth and Logic *(1936), which argued that "unverifiable statements—such as 'God exists,' 'human life has a distinct purpose,' or 'abortion is evil'—are scientifically mean-ingless. . . . They are pure opinion."* [3] *Succeeded JULIAN HUXLEY as pres-ident of the British Humanist Association. While teaching in the United States in 1987, Ayer, then 77, saw boxer Mike Tyson harassing model Naomi Campbell at a party, and demanded that he stop. "Do you know who the fuck I am?" Tyson asked. "I'm the heavyweight champion of the world." "And I am the former Wykeham Professor of Logic," Ayers replied. "We are both pre-eminent in our field. I suggest that we talk about this like rational men."*

"To say that 'God exists' is to make a metaphysical utterance which cannot be either true or false. . . . Not to confuse this view of religious assertions with the view that is adopted by atheists, or agnostics. . . . [Agnostics] hold that the existence of a god is a possibility in which there is no good reason either to believe or disbelieve; [atheists] hold that it is at least probable that no god exists. . . . Our view [is] that all utterances about the nature of God are nonsensical. . . . If the assertion that there is a god is nonsensical, then the atheist's assertion is that there is no god is equally nonsensical, since it is only a significant proposition that can be significantly contradicted. As for the agnostic . . . he does not deny that the two sentences 'There is a transcendent god' and 'There is no transcendent god' express propositions one of which is actually true and the other false. All he says is that we have no means of telling which of them is true, and therefore ought not to commit ourselves to either. But we have seen that the sentences in question do not express propositions at all. And this means that agnosticism also is ruled out." *Later variations on this position include "ignosticism" (see SHERWIN WINE) and "apathetic agnosticism" (see JOHN PARIURY).*

"The 'person' who is supposed to control the empirical world [but] is not himself located in it . . . is not an intelligible notion at all. We may have a word which is used [*the* G *word*], as if it named this 'person,' but . . . it cannot be said to symbolize anything. . . . The mere existence of the noun is enough to foster the illusion that there is a real, or at any rate a possible entity corresponding to it."

"None of those who have compared the world to a vast machine [made by the Great Watchmaker] has ever made any serious attempt to say what the machine could be for. . . . Theists have generally assumed that it had something to do with the emergence of man. This is a view which it is perhaps natural for men to take but hardly one that would be supported by a dispassionate consideration of the scientific evidence. Not only did man make a very late appearance upon the scene in a very small corner of the universe, but it is not even probable that, having made his appearance, he is here to stay. . . . So far as scientific evidence goes, the universe has crawled by slow degrees to a somewhat pitiful result on this earth, and is going to crawl by still more pitiful stages to a condition of universal death. If this is to be taken as evidence of purpose, I can only say that the purpose is one that does not appeal to me."

SECULAR * INFIDEL * NONBELIEVER * HUMANIST * RATIONALIST * FREETHINKER * AGNOSTIC * GODLESS * HERETIC * ATHEIST

Edward Babinski (1956–), *American librarian and "atheistic rabble-rouser." Former young-earth creationist and fundamentalist Christian. Editor of* Leaving the Fold: Testimonies of Former Fundamentalists *and of the periodicals* Monkey's Uncle *and* Cretinism or Evilution *[sic]. ("My spell-checker," he explained, "lacks the word 'creationism' in its dictionary, so each time that word is encountered, an alternative pops up at the bottom of my screen, 'cretinism.'")*

> "Don't creationists ever wonder about the fact that the paleontologists found *ape-like* skulls with the 'human leg and foot bones,' rather than the other way around, i.e., human skulls with 'ape leg and foot bones?' . . . Come on, creationists, think about it! Did God hide the human skulls, only leaving behind leg and foot bones belonging to *human midgets with misshapen feet,* and mix such bones only with the skulls of ape-like creatures with larger cranial capacities than living apes? What a 'kidder' the creationists' God must be."

Sir Francis Bacon (1561–1626), *English philosopher and statesman; father of the scientific method.* JOHN DRYDEN *wrote: "The World to Bacon does not only owe it's present knowledge, but its future too." Only 43 degrees of separation from Kevin Bacon.*

No one's saying he was an atheist. . . . The Bacon line that clerics love to quote: "A little philosophy inclineth man's mind to atheism, but depth in philosophy bringeth men's minds about to religion."

But in the very next essay: "Atheism leaves a man to sense, to philosophy, to natural piety, to laws, to reputation, all which may be guides to an outward moral virtue . . . but superstition dismounts all these, and erecteth an absolute monarchy in the minds of men."

And: "In every age, natural philosophy [science] had a troublesome adversary and hard to deal with; namely, superstition, and the blind and immoderate zeal of religion."

And: "If a man will begin with certainties, he shall end in doubts; but if he will be content to begin with doubts, he shall end in certainties."

Joe Bageant (1946–), *American writer/journalist. Vietnam veteran, hippie, Buddhist, and self-described "universalist humanist socialist." For seven years he lived and farmed off the grid on an Indian reservation in Idaho. Was friends with* TIMOTHY LEARY, *Allen Ginsburg,* WILLIAM BURROUGHS, *Marshall Mcluhan. With cred like that, he can say and write pretty much anything. And has. His articles have gained him "internet cult status." You can't buy ICS.*

"As the [2004] elections proved for once and for all, Christian fanatics . . . can no longer be written off as Dogpatch religionists. . . . It is one thing for them to have it in for their enemies, and quite another to have their own president, cabinet, Supreme Court, and newly established Department of Fatherland Surveillance backing them up."

"At the same time, the faithful presume themselves to be aggrieved holy victims, every last damned one of them. And when you are a victim, whether it be of the removal of the Ten Commandments from your white cracker court house by onanist liberal heathens 'frum up nawth,' or the refusal of the Great Satan Kansas school board to add humus and sheep's eyes to the school lunch program, you are entitled to revenge in the form of taking down the entire world. What the hell? God is gonna do it anyway at the end time. . . . About the only thing all three gods agree on is that exposed belly buttons and young folks having too much fun leads to the end of the world."

Julian Baggini, *British philosopher. Editor of* Philosophers' Magazine. *Author of* Atheism: A Very Short Introduction *(2003).*

"Atheism can be understood not simply as a denial of religion, but as a self-contained belief system . . . a commitment to the view that there is only one world and this is the world of nature."

"Goblins, hobbits . . . truly everlasting gobstoppers. . . . God is just one of the things that atheists don't believe in, it just happens to be the thing that, for historical reasons, gave them their name."

Kurt Baier (1917–), *professor emeritus of philosophy, University of Pittsburgh.*

"I suspect that many who reject the scientific outlook . . . confusedly think that if the scientific world picture is true, then their lives must be futile because . . . man has no purpose given him from without. These people mistakenly conclude that there can be no purpose *in* life because there is no purpose *of* life; that *men* cannot themselves adopt and achieve purposes. . . ."

Vanessa Baird, *British journalist; editor of the left-wing* New Internationalist *magazine. Author of* The No-Nonsense Guide to Sexual Diversity *and* The Little Book of Rebels—*one of those lazy, brain-candy collections of quotations from celebrities like* Christ *and* Gandhi.

"My mother had a medical attitude towards religion. If you didn't give children a good dose of it early on they might catch a more extreme case later in life."

From her 2004 article "In the name of God: are violence and religion natural bedfellows?" "Some 1,000 Catholic priests are under investigation in the U.S. alone on child sex abuse charges; 200 such cases are being investigated within Australia's

Anglican Church. Both churches have protected the perpetrators, rather than the victims, by simply moving offending clerics to other parishes."

Ibid.: "Faith is most robust in countries where there is great social inequality and poor state provision—in Nigeria, India and Indonesia over 90 percent of people count themselves as believers. This is no accident. . . . Where the state fails, religion steps in. . . . Even today, boys in remote parts of Afghanistan attend Taliban-run madrasas for the simple reason that they get a meal there. . . . Religious groups wield tremendous power over the uneducated and dependent poor."

"Today we are witnessing a new evangelical crusade coming from the West which has been dubbed 'evangelical capitalism.' This is more than laissez-faire economics: it sees 'the hand of God' in economic liberty, which in reality turns out to be the unfettered freedom of huge corporations to dominate national and global markets. The gospel according to Halliburton. Pitch this against the surge of Saudi-financed Wahabist fundamentalism imposing its all-conquering version of the only true Islam, and it's hard not to get trampled underfoot."

Carolyn Baker (ca. 1945–), *American professor of history and self-described "recovering fundamentalist Christian." Author of such articles as "The Religious Right: An Anti-American Terrorist Movement."*

"I recall my own dependency on what 'the Bible says.' . . . I remember the need for the 'fix' of the church service, the revival meeting, the prayer meeting. . . . But no 'fix' was more deliciously validating than 'winning souls for Christ'—that dramatic moment when I had manipulated someone else into a born-again experience. For this, the fundamentalist Christian addict lives and breathes."

"Born-again Christians worship the Bible and not God. . . . Bible worship is nothing less than 'having other gods before me. . . .'"

"The underlying [fundamentalist Christian] message is: 'You don't believe the Bible is the inerrant Word of God because your mind has been occupied by Satan.'"

"Christian fundamentalism in 'cafeteria style' has chosen which parts of Jesus' teachings it chooses to honor and which not. . . . Little attention is given to the Sermon on the Mount and the many passages where Jesus condemns the wealthy and the religious leaders of his time for their callous, hypocritical, mean-spirited absence of compassion. In fact, theologians who pay much attention to Jesus' teachings on compassion are viewed as bleeding hearts, unorthodox, and not really Christian." *(See* PAT ROBERTSON)

On the fundamentalist Christian "contempt for life":

"Being 'pro-birth' is not the same as being pro-life. Forcing females to have children without providing what they need financially, emotionally, and educationally is a pro-birth agenda that murders countless bodies and souls. . . . These individuals have an appalling disconnect, fawning over the decaying body of a woman in a permanent vegetative state [Terri Schiavo] while praising the demise of over 100,000 innocent Iraqi citizens and touting the patriotism of some 1,600 dead U.S. troops. . . . In [the religious right's] mindset, adult human lives do not matter because the human condition itself is inherently evil. . . ."

Robert A. Baker, *Emeritus professor of psychology, University of Kentucky, and noted spiritual bunk debunker. Contributor to* Skeptical Briefs, *the newsletter of the Committee for the Scientific Investigation of Claims of the Paranormal.*

"What happens when the same number of people pray for something as pray against it? How does God decide whose prayer to answer? . . . This spring when a small Kentucky town won the State High School Girl's Basketball crown, the town's newspaper, as well as the largest newspaper in Kentucky, gave credit for the victory to God's answering their prayers. Why their prayers were answered and the prayers of the losers were not remains unknown. One possibility is that the Hazard team had a better 'pray-er'—in the form of their principal, who was also a minister. If it turns out that the higher one stands in the religious hierarchy the better the chances that one's prayers will be heeded, then it certainly behooves every athlete and every athletic team to employ the most religious 'pray-ers' possible. Certainly no one should ever enter any contest unpre-prayered!"

Joan Bakewell (1933–), *British journalist and TV host since the 1960s. Her 2004 autobiography described her affair with* HAROLD PINTER *while both were married. (There you are: without religion. . . .)*

On the murder of 24-year-old TV and radio presenter Shaima Rezayee in liberated Afghanistan in 2005, after Muslim clerics attacked her MTV like show ("it will corrupt our society . . . take our people away from Islam and destroy our country") and Western style of dress:

> "Rezayee was at the crossroads of a punitive tradition that fears and resents women and the new tradition . . . that celebrates them. . . . The control of dress might seem a petty matter, but it is loaded with significance."

> "But why are religions so tough on women? In the Victorian heyday of muscular Christianity, the rules of feminine dress would have met the highest standards of the Qur'an. It was in religiously devout America that Janet Jackson's breast caused so much fuss. Only as we have become more secular have we shed our clothes and our inhibitions. Who are these gods that they should require their own creatures to be ashamed of their bodies? . . . The notion that the supposed creator is offended by the natural beauty of his own creation is well nigh blasphemous."

Tammy Faye Bakker Messner (1942–), *a.k.a. Our Lady of Mascara: Christian TV personality. Ex-wife of convicted felon/televangelist Jim Bakker, with whom she cohosted the* PTL [Praise The Lord] *Club.*

> "I take Him shopping with me. I say, 'OK, Jesus, help me find a bargain.'"

Mikhail Bakunin (1814–1876), *Russian anarchist leader/theorist/writer. Split from the Marxists over his opposition to "authoritarian socialism," saying: "If you took the most ardent revolutionary, vested him in absolute power, within a year he would be worse than the Czar himself." ("Two legs bad," I always say.)*

"We are materialists and atheists, and we glory in the fact."

"The first revolt is against the supreme tyranny of theology, of the phantom of God. As long as we have a master in heaven, we will be slaves on earth."

"The idea of God implies the abdication of human reason and justice; it is the most decisive negation of human liberty."

"People go to church for the same reasons they go to a tavern: to stupefy themselves, to forget their misery, to imagine themselves, for a few minutes anyway, free and happy." *(A tavern, or an opium-of-the-people den.)*

"Does it follow that I reject all authority? Perish the thought. In the matter of boots, I defer to the authority of the bootmaker."

James Baldwin *(1924–1987), African-American writer. The Harlem-born Baldwin, much of whose work dealt with being black and homosexual in mid-twentieth century America, once said: "All of Africa will be free before we can get a lousy cup of coffee."*

"If the concept of God has any validity or any use, it can only be to make us larger, freer, and more loving. If God cannot do this, then it is time we got rid of Him."

All due respect to VOLTAIRE: "If God *existed,* it would be necessary to abolish him."

Honoré de Balzac *(1799–1850), French novelist; a founder of the realist school of fiction.*

Such a Romantic: "After a woman gets too old to be attractive to men, she turns to God." (Compare MADAME DE STAËL.)

Iain M. Banks *(1954–) Scottish science fiction and fiction-fiction writer. "[W]rites novels with plots so fantastic they almost make the Bible sound plausible," wrote one interviewer. "To some, his books are sadistic, evil and sick." Writes a novel a year, in three months—takes the other nine off to relax and replenish his depravity. Calls himself an "evangelical atheist."*

Member of a group of prominent Britons who campaigned in 2004 to have Prime Minister Tony Blair impeached for the invasion of Iraq. Cut up his passport in protest and mailed it to 10 Downing Street.

"Faith is wrong; belief without reason and question is evil."

On the appeal of cults (he invented one in his novel Whit*):*

"It's the same as the appeal of joining the army, or joining any highly-disciplined organisation that takes away choice. The more sophisticated and complex society gets . . . the more confused you can get. . . . People crave certainty even if it's a specious certainty."

"Cults and sects and religions tend to be set up by men because they're a power trip. . . . Look at David Koresh of Waco fame. He tried to be a rock star and failed. As a prophet though, he got the rock star life, the sex and drugs and worship, without having to be one. . . !" *In* Whit, *a female character suggests religions are mainly devised by men because males suffer from "ovary envy"— envy of women's power of creation.*

"'Put your trust in the Lord,' goes their always unspoken motto, 'your ass belongs to us.'"

Reverend Sabine Baring-Gould *(1834–1924), English parson, antiquarian, and writer of hymns including "Onward, Christian Soldiers." Author of* The Book of Were-Wolves *(1865), a frequently cited work on the subject. Keeper of a pet bat. At age 34, married a mill girl of 16. The marriage—said to be the basis of* SHAW's Pygmalion*—lasted 48 years and produced 15 children. At a children's party once, he asked: "And whose little girl are you?" The child burst into tears and said, "I'm yours, Daddy."*

"The religious passion verges so closely on the sexual passion that a slight additional pressure given to it bursts the partition, and both are confused in a frenzy of religious debauch."

Dan Barker *(1949–), American atheist writer and activist; copresident, Freedom From Religion Foundation (FFRF). Former evangelical preacher; maintained a touring musical evangelical ministry for 17 years.*

Author of Losing Faith in Faith: From Preacher to Atheist*; Just Pretend: A Freethought Book for Children*; *and* Maybe Yes, Maybe No: A Guide for Young Skeptics. *Member of the Lenni Lenape Tribe of Native Americans.*

"Truth does not demand belief. Scientists do not join hands every Sunday, singing, 'yes, gravity is real! I will have faith! I will be strong! I believe in my heart that what goes up, up, up must come down. . . . Amen!' If they did, we would think they were pretty insecure about it."

"You believe in a book that has talking animals, wizards, witches, demons, sticks turning into snakes, food falling from the sky, people walking on water, and all sorts of magical, absurd and primitive stories, and you say that *we* are the ones that need help?"

"I [as a believer] assumed that the successful prayers were proof that God answers prayer while the failures were proof that there was something wrong with me."

"We think the National Day of Prayer is unconstitutional. What if the president declared a National Day of Cursing God because He failed us on September 11? . . . That's how we feel when he promotes prayer."

"The very concept of sin comes from the Bible. Christianity offers to solve a problem of its own making! Would you be thankful to a person who cut you with a knife in order to sell you a bandage?"

Ronald J. Barrier*, National Spokesperson, American Atheists. On a survey showing higher divorce rates among Jews and Christians, especially born-again Christians, than among atheists and agnostics:*

"Since Atheist ethics are of a higher caliber than religious morals, it stands to reason that our families would be dedicated more to each other than to some invisible monitor in the sky. With Atheism, women and men are equally responsible for a healthy marriage. . . . Atheists reject, and rightly so, the primitive patriarchal attitudes so prevalent in many religions with respect to marriage." *We* spit *on primitive patriarchal attitudes.*

Dave Barry (1947–) *American humorist. Son of a Presbyterian minister. Elected class clown in high school. Avoided military service during the Vietnam War by registering as a religious conscientious objector, even though, as he has said, "I decided I was an atheist early on." Won a Pulitzer Prize for Commentary in 1988.*

"The problem with writing about religion is that you run the risk of offending sincerely religious people, and then they come after you with machetes." *

"As far as I could tell, there's nothing preachy about Buddhism. I was in a lot of temples [in Japan], and I still don't know what Buddhists believe, except that at one point Kunio [?] said 'If you do bad things, you will be reborn as an ox.' This makes as much sense to me as anything I ever heard from, for example, the Reverend Pat Robertson." *(May he come back as a sea slug.)*

"In fact, when you get right down to it, almost every explanation Man came up with for *anything* until about 1926 was stupid."

Bruce Bartlett (1951–), *American conservative economist and commentator. Domestic policy adviser to President Reagan; treasury official under George H. W. Bush. Author of* Impostor: How George W. Bush Bankrupted America and Betrayed the Reagan Legacy. *His criticism of Bush got him fired in 2005 from the free-market think tank he'd been affiliated with since 1993.*

October 2004: "Just in the past few months, I think a light has gone off for people who've spent time up close to Bush: that this instinct he's always talking about is this sort of weird, Messianic idea of what he thinks God has told him to do. This is why George W. Bush is so clear-eyed about Al Qaeda and the Islamic fundamentalist enemy. . . . He understands them, because he's just like them. This is why he dispenses with people who confront him with inconvenient facts, he truly believes he's on a mission from God. Absolute faith like that overwhelms a need for analysis. . . . But you can't run the world on faith."

---

*Attacks with axes have also been reported.

Charles Baudelaire (1821–1867), *French poet and critic. He, his publisher, and the printer of his first and most famous book of poems,* The Flowers of Evil, *were successfully prosecuted for offending public morals.*

"God is the only being who, in order to reign, doesn't even need to exist."

"What matters an eternity of damnation to someone who has found in one second the Infinity of joy?"

"Unable to do away with love, the Church found a way to decontaminate it by creating marriage."

"I have always been astonished that women are allowed to enter churches. [*What?? Since when?*] What talk can they have with God?"

Alfred-Henri-Marie Cardinal Baudrillart (1859–1942), *Roman Catholic priest. Auxiliary Bishop of Paris, 1921–1942.*

"Hitler's war is a noble undertaking in defense of European culture." *(1941)*

Pierre Bayle (1647–1706), *French philosopher and literary critic. Son of a Calvinist minister. Fled to Switzerland to avoid persecution of Protestants. Argued that religion and morality are two quite, quite different things.*

"In matters of religion it is very easy to deceive a man, and very hard to undeceive him."

On the appearance of Halley's Comet in 1680, which clerics said portended God's wrath:

"To offer such explanations in seriousness, shows the greatest contempt of mankind. . . . [Using natural phenomena] as marks of the wrath of heaven [is] to the interest of pontiffs, priests, and augurs, as much as it is to the interest of lawyers and doctors that there should be lawsuits and sickness."

BBC Online *American religious looniness faces growing foreign competition.*

August 2002: "Jedi 'religion' grows in Australia—More than 70,000 people in Australia have declared that they are followers of the Jedi faith, the religion created by the Star Wars films. A recent census found that one in 270 respondents—or 0.37% of the population—say they believe in 'the force,' an energy field that gives Jedi Knights like Luke Skywalker their power in the films."

Simone de Beauvoir *(1908–1986), French author, philosopher, feminist. Long-time companion and doormat of JEAN-PAUL SARTRE. Also had a lesbian relationship with one of her students (who meanwhile avait des relations with Sartre). Didn't have her first full orgasm until age 39, when American writer Nelson Algren came into her life. She told all about it in subsequent books. He wasn't pleased. (How do you suppose Sartre felt?)*

Reading BALZAC one night: "'I no longer believe in God,' I told myself, with no great surprise. . . . That was proof: if I had believed in Him, I should not have allowed myself to offend Him so light-heartedly. I had always thought that the world was a small price to pay for eternity; but it was worth more than that, because I loved the world, and it was suddenly God whose price was small: from now on His name would have to be a cover for nothing more than a mirage."

"Since man exercises a sovereign authority over woman, it is especially fortunate that this authority has been vested in him by the Supreme Being."

"The aversion of Christianity in the matter of the feminine body is such that while it is willing to doom its God to an ignominious death, it spares him *[by means of the Virgin Birth]* the defilement of being born."

Referring to the Council of Nicaea, 325 C.E., the first "international" conference of Christian bishops, which, by a single vote, declared woman to be "human":

"Christianity gave eroticism its savor of sin and legend when it endowed the human female with a soul."

Samuel Beckett (1906–1989), *Irish playwright, novelist, poet. Working with the French Resistance during World War II, he narrowly escaped capture after his unit was betrayed to the Gestapo by a former Catholic priest.*

The character Hamm in Beckett's play Endgame, *after attempting to pray:* "The bastard! He doesn't exist!"

John Leonard Beevers (1911–1975), *English journalist and historian. Author of* World without Faith *(1935), a defense of freethought. Later "converted" from Communism to Roman Catholicism and turned to writing biographies of Catholic saints and translations of Catholic theological treatises. Win some, lose some.*

"I do not know that Christianity holds anything more of importance for the world. It is finished, played out. The only trouble lies in how to get rid of the body before it begins to smell too much."

Aphra Behn (1640–1689), *English novelist and dramatist. Said to be the first woman to support herself by writing. Spied for England during the Second Anglo-Dutch War by seducing a Dutch royal. Reportedly bisexual (making her perhaps a more versatile spy). Described her life as "dedicated to pleasure and poetry."*

"The gods, by teaching us religion, first set the world at odds."

"There is no sinner like a young saint." (AUGUSTINE *being the classic case.*)

Francis Bellamy (1855–1931), *American Baptist minister; author of the original Pledge of Allegiance (1892). His socialist convictions cost him his Boston pastorate in 1891.*

Scott Bellamy, great-grandson: "You'd think he would not have had bad feelings about having 'under God' [*added by Congress in 1954*] in the Pledge. But he was not even happy about them adding 'to the United States of America'" *(which Francis called a "clumsy redundancy . . . a mangling of the original").*

Great-granddaugher Sally Wright: "As a regular churchgoer who has voted both Democratic and Republican, I believe that my great-grandfather got it right. A Pledge of Allegiance that does not include God invites the participation of more Americans."

St. Robert Bellarmine (1542–1621), *Italian cardinal.*

"The Pope may act outside the law, above the law, and against the law."

"Freedom of belief is pernicious. It is nothing but the freedom to be wrong."

Catherine Bennett, *British journalist. Columnist for the left-liberal Guardian. October 2001—on the Religious and Racial Hatred Bill proposed by the Blair government after 9/11 and enacted in 2006, which in its original form would have criminalized even "insulting" language:*

"At last, 12 years after they first burned copies of the Satanic Verses, RUSHDIE's fiercest opponents finally have a chance to ban the book in Britain for ever. Maybe Rushdie, his publishers and distributors will end up with seven years in prison! . . . [But the law] may not, in practice, even be much of a blessing to the Muslims it is designed to protect. They also enjoy freedom of expression. Until quite recently, more colourful Muslim enthusiasts such as [formerly British-based cleric] Omar Bakri Mohammed . . . joyfully exercised that freedom, calling for, among other things, a holy war in Britain. Won't they, too, miss it when it's gone?"

Steve Benson, *Pulitzer Prize–winning editorial cartoonist for the Arizona Republic. Grandson of former U.S. Secretary of Agriculture and*

LDS (Mormon) church president Ezra Taft Benson. From a 1997 essay titled "Goodbye to God":

> "To understand why I jumped from the Mormon wagon train requires an understanding of what Mormons are and how they think. . . . They really aren't much different from millions of poor, guilt-ridden souls who, throughout the march of human history, have hitched their hopes to mass movements of one sort or another. [*Quoting* ERIC HOFFER:] 'A rising mass movement attracts and holds a following by the refuge it offers from the anxieties, barrenness and meaninglessness of an individual existence.' . . . Once I realized this, it wasn't much of a leap out of religion altogether."

Jeremy Bentham *(1748–1832), English philosopher and social reformer. Advocated church-state sep, abolition of slavery and corporal punishment, equal rights for women, inheritance taxes . . . all that good stuff. Professed atheism in a book written under a pseudonym. As requested in his will, when he died his organs were removed for medical research—in contravention of the religious-based law against dissection; and, to thumb his posthumous nose at Christian burial, his body was preserved—fully dressed and sitting in a chair—and stored in a wooden cabinet, termed his "Auto-Icon," at University College London, which he helped establish. It has been brought out of storage for council meetings at which Bentham is listed on the roll as "present but not voting."*

> "There is no pestilence in a state like a zeal for religion, independent of morality."

> "The spirit of dogmatic theology poisons everything it touches."

Nikolai Berdyaev *(1874–1948), Russian philosopher. A Christian who was sentenced to exile in Siberia for life for a 1913 article criticizing the Russian Orthodox Church, but was saved by the Russian Revolution—then deported by the Bolshies in 1922.*

> "We find the most terrible form of atheism, not in the militant and passionate struggle against the idea of God himself, but in

the practical atheism of everyday living, in indifference and torpor. We often encounter these forms of atheism among those who are formally Christians."

Bernard Berenson (born Bernhard Valvrojenski, 1865–1959),

Lithuanian-Jewish born American art historian. Considered the world's leading Renaissance art connoisseur in his day. Art critic for The Nation *magazine.*

"Miracles happen to those who believe in them. Otherwise why does not the Virgin Mary appear to Lamaists, Mohammedans, or Hindus who have never heard of her?"

José Bergamín (1895–1983), *Spanish writer. Headed the Alliance*

of Antifascist Intellectuals during the Spanish Civil War. His mother was a fervent Catholic and *communist who once said, "With Communism until death—but not a step further."*

"You need to have a God, a lover, and an enemy, says the poet. Exactly: you need to have three enemies."

Ingmar Bergman (1918–), *Swedish film and theater director.*

Son of a Lutheran minister. God, mortality, and his own gradual shift from belief to nonbelief are central themes of his films. As are hot blond babes. With tormented souls.

"I hope I never get so old I get religious."

"By mistake I was given too much anesthesia [before a minor surgery]. The lost hours of that operation provided me with a calming message. You were born without purpose, you live without meaning, living is its own meaning. When you die, you are extinguished. From being you will be transformed to non-being. . . . This insight has brought with it a certain security that has resolutely eliminated anguish and tumult. . . ." *Reportedly told an interviewer in 2000 he believes in supernatural worlds, communicates with his dead wife, and is looking forward to reuniting with her in the next world.*[4]

Sir Isaiah Berlin (1909–1997), *Latvian-born British-Jewish political philosopher and intellectual historian. Winston Churchill once invited songwriter Irving Berlin to lunch, having confused him with Isaiah Berlin.*

> "As for the meaning of life, I do not believe that it has any . . . and this is a source of great comfort to me. . . . Those who seek for some cosmic all-embracing libretto or God are, believe me, pathetically mistaken."

Andrew Bernstein, *American philosopher and novelist. Disciple of AYN RAND. Author of a novel,* Heart of a Pagan, *whose Randian hero is a superathlete, a basketball player, sent by the hoops-loving gods of ancient Greece to inspire the worship of mental, physical, and moral excellence. No, I'm not kidding.**

> "The Platonic-Christian tradition in philosophy trumpets two claims: (1) that man is a being severed into two parts, that his body belongs to this dimension of reality and his consciousness to a higher, spiritual realm—and (2) the logical consequence of this mind-body split, the belief that this world is utterly material and carnal [and] that the intellect, since it belongs to another world, is helpless to deal with this one. . . . Just as Jesus is the perfect moral expression of this view—the weak, pacifistic, cheek-turning 'lamb' in this world, but the omnipotent deity ruling the next—so Hamlet is its perfect literary expression—the brilliant philosopher-intellectual who excels in the theoretical realm but is helpless to deal with the practical."

Matt Berry, *American philosopher and writer. Author of* Post-Atheism: A Mechanist's Journey from Christian Materialism to Material Spirituality *(2001).*

> "Faith is the fatigue resulting from the attempt to preserve God's integrity instead of one's own."

> "Atheism draws me out of the herd of theism. Atheism is in this sense a necessary point of arrival. It is how and where I *stop* my

---

* Except for the "hoops-loving gods."

inherited cultural inertia. . . . [But atheism] too is a herd I leave behind, for this is not a God-AntiGod reality. . . . This is *post*-atheism. . . . I accept the predicament: 'I am a machine, and my function is to lie to myself.' . . . A new spirituality after atheism begins here."

Annie Wood Besant (1847–1933), *English philosopher-activist. Was married off at age 19 to a preacher who eventually kicked her out over her growing irreligiousness. (He got the children, of course.) Organized women factory workers. Campaigned for women's suffrage. Elected to the London School Board, she instated free meals for poor students and free medical exams for every pupil. She and* CHARLES BRADLAUGH *coauthored* The Freethinker's Textbook *and were co-arrested for co-promoting birth control. Wrote* The Gospel of Atheism *(1877) and* Why I Do Not Believe in God *(1887). Later got into Hinduism, Buddhism, yoga, and, going for broke, Theosophism.*

"An Atheist is one of the grandest titles . . . it is the Order of Merit of the World's heroes . . . Copernicus, Spinoza, Voltaire, Paine, Priestly."

"For centuries the leaders of Christian thought spoke of women as a necessary evil, and the greatest saints of the Church are those who despise women the most. . . . This coarse and insulting way of regarding woman, as though they existed merely to be the safety-valves of men's passions, and that the best men were above the temptation of loving them, has been the source of unnumbered evils."

Buddhadasa Bhikkhu (1906–1993), *Thai Buddhist monk and reformer. Inspired Thailand's 1932 revolution and 1960s social activism. His books, which include* Handbook for Mankind *and* No Religion, *literally take up an entire room in the National Library of Thailand.*

"Even the present life does not exist. How could the after-life exist?"

John Bice, *American writer, religion columnist, and noted beer expert.*

"As bumper-sticker philosophy points out, 'religions are just cults with more members.' . . . What's the difference, rationally speaking, between believing [*as Scientologists reportedly do*] in body-infesting souls and ancient galactic confederations, or in the stories of virgin birth, Vishnu, the Garden of Eden, transubstantiation, Noah's ark, Judgment Day, or the baseless concept of the Trinity?"

"The vast majority of personal religious beliefs can be accurately predicted based solely on the beliefs of one's parents or the culture one is raised in. . . . Religionists should ask themselves, 'Are my religious beliefs based on rationality and evidence or indoctrination?'"

"A belief in an afterlife has the unavoidable effect of making this life less unique and precious. . . . Good luck finding an atheist willing to strap a bomb to his or her back, or fly a plane into a building. . . ."

Ambrose Bierce (1842–1914), *American satirist, poet, and critic. Author of* The Devil's Dictionary, *originally a newspaper serialization published between 1881 and 1906. Whence:*

"*Christian:* One who follows the teachings of Christ insofar as they are not inconsistent with a life of sin."

"*Clairvoyant:* A person, commonly a woman, who has the power of seeing that which is invisible to her patron—namely, that he is a blockhead."

"*Clergyman:* A man who undertakes the management of our spiritual affairs as a method of bettering his temporal ones."

"*Evangelist:* A bearer of good tidings, particularly such as assure us of our own salvation and the damnation of our neighbours."

"*Infidel:* In New York, one who does not believe in the Christian religion; in Constantinople, one who does."

"*Ocean:* A body of water occupying about two thirds of a world made for man—who has no gills."

"*Pray:* To ask that the laws of the universe be *annulled* in behalf of a single petitioner confessedly unworthy."

"*Rack:* An argumentative implement formerly much used in persuading devotees of a false faith to embrace the living truth."

"*Religion:* A daughter of Hope and Fear, explaining to Ignorance the nature of the Unknowable."

"*Reverence:* The spiritual attitude of a man to a god and a dog to a man."

"*Saint:* A dead sinner revised and edited."

"*Scriptures:* The sacred books of our holy religion, as distinguished from the false and profane writings on which all other faiths are based."

Mark K. Bilbo (1961–), *host and operator of alt-atheism.org, the official site of the alt.atheism Usenet newsgroup.*

"The very need for a thing called 'apologetics' is example of the weakness of the theistic argument. 'God' always needs apologies, rationalizations, explanations, equivocations, excuses."

Arthur "Pitcher" Binstead (1846–1915), *British journalist, founder of London's Pelican Club, and the inspiration for at least two characters in Arthur Conan Doyle's Sherlock Holmes stories.*

"The most serious doubt that has been thrown on the authenticity of the biblical miracles is the fact that most of the witnesses in regard to them were fishermen."

Björk (Björk Guðmundsdóttir, 1965–), *Icelandic singer/songwriter. Made her first album at age 11. At 14, formed an all-girl punk band, Spit and Snot. At 16, formed another band, Tappi Tíkarrass ("Cork the Bitch's Ass"). Hers, one could already see, would be a life of piety, purity, and respect for traditional values. In Icelandic, Björk rhymes with "jerk."*

"I've got my own religion. . . . Iceland sets a world-record. . . .
When we were asked [on a worldwide UN survey] what do we
believe, ninety percent said, 'ourselves.' I think I'm in that
group."

"The Buddhists say we come back as animals and they refer to
them as lesser beings. Well, animals aren't lesser beings, they're
just like us. So I say fuck the Buddhists."

Lewis Black (1948–), *American comedian. Best known for his commentary segment on* The Daily Show, *his ranting, verge-of-a-nervous-breakdown style, and his filthy, disgusting language. Introduced as an "angry agnostic" on Comedy Central's* Bar Mitzvah Bash.

"You know, that's why our enemy is so frightening—they have
no humor. This is a group of people who wandered the desert for
thousands and thousands of years and never ran into a knock-
knock joke." *(Also see AYATOLLAH KHOMEINI.)*

William Blake (1757–1827), *English Romantic poet, artist, publisher, mystic.*

"I care not whether man is good or evil; all that I care / Is
whether he is a wise man or a fool. Go, put off holiness / And
put on intellect . . ."

"As the caterpillar chooses the fairest leaves to lay her eggs on,
so the priest lays his curse on the fairest joys."

"Prisons are built with stones of Law, Brothels with bricks of
Religion."

From his poem "The Divine Image," which describes God "as the imagined embodiment or incarnation of values generated by the wholly human spirit" [15]:

"To Mercy, Pity, Peace, and Love / All pray in their distress; / And
to these virtues of delight / Return their thankfulness. / For
Mercy has a human heart, / Pity a human face, / And Love, the
human form divine, / And Peace, the human dress. / Then every
man, of every clime, / That prays in his distress, / Prays to the
human form divine, / Love, Mercy, Pity, Peace."

David Bloomberg, *chairman and cofounder of the Rational Examination Association of Lincoln Land (Illinois), and* Steven Novella, M.D., *president and cofounder of the New England Skeptical Society.*

> "To a scientific rationalist, there is no distinction between believing in leprechauns, alien abductions, ESP, reincarnation, or the existence of a God—each equally lacks objective evidence. . . . Separating out the latter two beliefs and labeling them as religion—thereby exempting them from critical analysis—is intellectually dishonest. . . . [Indeed] the most widespread and sacredly guarded superstitions [are] the most important ones to oppose, for they have the greatest influence and can therefore do the most harm."

Simon Bolivar *(1783–1830), "El Libertador" of much of South America from Spanish rule. Excommunicated by the Catholic Church for his professed atheism. (They do get upset over the smallest things.)*

> "Should the State rule the conscience of its subjects? Watch over the fulfilment of religious laws? . . . Was a modern State to restore the Inquisition and burnings at the stake?"

Napoleon Bonaparte *(1769–1821), French general and emperor. Avid Jacobin and anticlerical during the Revolution. After conquering Egypt, he declared himself a Muslim to gain local support, but upon his return to France declared: "Among the Turks, I was a Mohammedan. Now I shall become a Catholic." After conquering Italy, he imprisoned the Pope and demanded His Holiness submit to his authority. (HH had him excommunicated. Boney laughed: "In these enlightened days, none but children and nursemaids are afraid of curses.") At 5 feet 6.5 inches, he was slightly taller than the average Frenchman of his day.*

> "Everything is more or less organized matter. To think so is against religion, but I think so just the same."

> "A soul? Give my watch to a savage, and he will think it has a soul."

> "If I had believed in a God of rewards and punishments, I might have lost courage in battle."

"God fights on the side with the best artillery."

"As for myself, I do not believe that such a person as Jesus Christ ever existed; but as the people are inclined to superstition, it is proper not to oppose them."

"My firm conviction is that Jesus . . . was put to death like any other fanatic who professed to be a prophet or a messiah; there have been such persons at all times. . . . Besides, how could I accept a religion which would damn Socrates and Plato?"

"I am surrounded by priests who repeat incessantly that their kingdom is not of this world, and yet they lay their hands on everything they can get."

"How can you have order in a state without religion? For, when one man is dying of hunger near another who is sick from overeating, he cannot resign himself to this difference unless there is an authority which declares, 'This is God's will.'. . . . Religion is what keeps the poor from murdering the rich."

Sir Hermann Bondi (1919–2005), *Austrian-Jewish-British mathematician and cosmologist. Important contributor to general relativity theory. Self-described humanist who never "felt the need for religion" and for whom arguments about "the existence or nonexistence of an undefined God are quite pointless."*

"Every one of us . . . has met the criticism that in ethics we humanists live on Christian capital, that our moral attitudes are derived from Christianity. I believe this to be utterly wrong and that, on the contrary, what goes for modern Christian ethics is in fact derived from humanist values. For most of its history Christianity was red in tooth and claw. . . . It is only in the last couple of centuries that Christian attitudes have gradually become 'civilised' and humane. Why? [Because of] the rise of humanism and skepticism. We have given Christianity its modern face, which often quotes the very nice things Jesus is reported to have said, and carefully omits the nasty sayings such as 'If a man abide not in me, he is cast forth as a branch, and is withered; and men gather them, and cast them into the fire, and they are burned.'"

Hypatia Bradlaugh Bonner (1858–1934), *English feminist and peace activist, death penalty opponent, and Justice of the Peace for London (1922–1934). Daughter of* CHARLES BRADLAUGH, *whose affairs she took over when he was arrested. Successfully sued someone who alleged her father had had a deathbed conversion. Named after Hypatia of Alexandria, a fourth-century pagan philosopher who was hacked to death in a church, during Lent, by a Christian mob led by a preacher and urged on by Saint Cyril, apostle to the Slavs and creator of the Cyrillic alphabet.*

Motto of Bonner's journal The Reformer: "Heresy makes for progress."

"Before August, 1914, it was the correct thing to proclaim Christ as the Prince of Peace and Christianity as the religion of love. . . . Lip-service was paid to Peace from thousands of pulpits. After August, 1914, these same pulpits resounded with praises of the Lord as a man of war (Exodus, xv. 3) and declarations that the great European War was a Christian war, sent directly by Almighty God himself."

"Now, in my seventy-eighth year, being of sane mind, I declare without reserve or hesitation that I have no belief, and never have had any belief, in any of the religions which obsess and oppress the minds of millions of more or less unthinking people throughout the world."

Bono (Paul David Hewson, 1960–), *Irish rock star and international do-gooder. Catholic? Protestant? "I always felt like I was sitting on the fence." In 2005 he gave George W. Bush an iPod and the book* The Message, *a controversial translation of the Bible into contemporary language. Like Bush needed encouragement.*

"These are big questions. If there is a God, it's serious. And if there isn't a God, it's even more serious. Or is it the other way around? I don't know, but these are the things that, as an artist, are going to cross your mind—as well as 'Ode to My New Jaguar.'"

"I often wonder if religion is the enemy of God. It's almost like religion is what happens when the Spirit has left the building." *Okay, we did a rock star.*

Daniel J. Boorstin (1914–2004), *American historian and Librarian of Congress. Pulitzer Prize winner, 1973. His 1962 book* The Image: A Guide to Pseudo-Events in America *explored the relationship between reality and advertising, media, and simulation—influences that perhaps help explain our descent into a faith-based, deception-based society.*

> "The world has suffered far less from ignorance than from pretensions to knowledge. . . . No agnostic ever burned anyone at the stake or tortured a pagan, a heretic, or an unbeliever."

> "God is the Celebrity-Author of the World's Best Seller. We have made God into the biggest celebrity of all, to contain our own emptiness."

Elayne Boosler (1952–), *American comedian. According to her Web site, "Her sole goal in college was to turn eighteen so she could legally leave school and move to Manhattan to fulfill her dreams of waitressing."*

> "The Vatican is against surrogate mothers. Good thing they didn't have that rule when Jesus was born."

Jorge Luis Borges (1899–1986), *Argentine short story writer, poet, essayist, and critic. An avant-gardist from teen age, he began seriously fucking with readers' heads around 1933, freely mixing nonfiction, fiction, the fantastic, literary forgeries, and reviews of imaginary works, and exploring philosophical, mythological, mathematical, and religious themes. Self-described "*Spencerian *anarchist" or "Anarcho-Pacifist." One suspects he regarded such terms as "theist," "atheist" and even "agnostic" as idiotically simplistic, self-certain and boring: This dude questioned his own existence and that of the universe. Glaucoma gradually destroyed his vision. Never learned Braille. Lived with his mother, who looked after him and served as his personal secretary, until her death at age 99.*

> *Upon becoming completely blind around age 55:* "Let no one lower to tears or reproach / This declaration of God's Mastery, / That with magnificent irony / Gave me books and night at once."

> *From his story "Hakim, the Masked Dyer of Merv":* "The earth

we inhabit is an error, an incompetent parody. Mirrors and paternity are abominable because they multiply and affirm it."

"We (the indivisible divinity that works in us) have dreamed the world . . . but we have allowed slight, and eternal, bits of the irrational to form part of its architecture so as to know that it is false."

"Doubt is one of the names of intelligence."

A character in his story "Death and the Compass": "[Christianity] belongs to the history of Jewish superstitions."

"Heaven and hell seem out of proportion to me: the actions of men do not deserve so much."

"To die for a religion is easier than to live it absolutely." *(Equating the two, some recent Islamists have accomplished both.)*

"Life itself is a quotation."

Max Born *(1882–1970), German-Jewish-British mathematician. Nobel Prize in Physics. Maternal grandfather of singer-actress Olivia Newton-John. It was to Born that* EINSTEIN *said of quantum mechanics, "God does not play dice."*

"Science is so greatly opposed to history and tradition that it cannot be absorbed by our civilization."

David Boulton, *British writer and antiwar activist, one-time boy-evangelist, failed politician, former Head of News, Current Affairs, and Religion at Granada TV, Britain's largest independent TV company. Radical religious (Quaker) humanist. Author of* The Trouble with God *(2003). From* The New Internationalist *magazine, 2004:*

"There was a time, beginning around the 1850s and culminating perhaps in the 1920s, when it really did seem that the jig was up for organized religion—at least in the Western world. . . . [But] the growing complexities and insecurities of the 20th century paved the way for a triumphal return of the old certainties, promises, reassurances. God was resurrected. Today, 20 million grown-up Americans and 33 per cent of the Republican Party believe the

Rapture is imminent, when Christ will return to allow born-again evangelicals to share with him in divine governance of the universe. Hollywood finds the flagellation of Jesus a bigger turn-on than the female orgasm. The Rapture books in the Left Behind series . . . outsell Harry Potter."

"In Britain, the churches continue to empty, but the 'mind/body/spirit' shelves in our bookshops groan under the weight of tomes recommending a thousand varieties of bottled spiritualities. . . . One in ten men and one in four women tell pollsters they think there's something in reincarnation. One in three women say they believe in angels, particularly the guardian variety.

"It once seemed that reason was leading us to lose faith in religion, but we woke up to find, instead, that we had lost faith in reason. So all the old appurtenances of religion which we had chucked out through the doorway came creeping back through the window."

"Is religion, then, inevitable? Do we need it, as we need food, drink and sex? Do we, after all, have some kind of god-shaped gene. . . . Are we made with a religious itch which we must scratch. . . ? After all, religious belief and practice seems to have been part and parcel of virtually every human culture from the Neanderthals onwards. . . . Can it be just one long mistake? Was the whole of humanity on the wrong track from the year dot till the formation of the Rationalist Press Association?"

"We blind ourselves to the irrationalism, the bigotry, the fantasy of it all [because] religion gives us suitably solemn funerals, suitably sentimental nativity plays, provides us with life markers. It gives us our roots and our reassurance that there is meaning, even if it is located above the bright blue sky and we don't have a clue what the meaning means."

"Only a blinkered, anorexic humanism chooses to ignore the heritage of religious culture. . . . We still need a little salve-ation, healing, from time to time; a sense of at-one-ment with ourselves and the rest of the universe; redemption as restoration; an assurance that our ludicrous inability to be the people we would like to be is ultimately forgivable and forgiven."

Gerardus Bouw (1945–), *Dutch-American astronomer. Heads the Association for Biblical Astronomy. Leading figure of "modern geocentrism": Yes, insists that, as several Bible passages imply, the Earth is the center of the universe and does not move. His 1992 book* Geocentricity *has been described as "the most sophisticated defence of geocentricity ever published." (Not saying very much? Probably.)*

> *Here's how we know the universe revolves around the eartth:*
> "Historians readily acknowledge that the Copernican Revolution spawned the bloody French and Bolshevic [*sic*] revolutions . . . set the stage for the [revival of the] ancient Greek dogma of evolution . . . led to Marxism and Communism. . . . It is thus a small step to total rejection of the Bible and the precepts of morality and law taught therein."

> "If God cannot be taken literally when He writes [in the Bible] of the rising sun, then how can one insist that he be taken literally when writing of the rising of the Son?"

David Bowie (born David Robert Jones, 1947–), *English pop singer-songwriter. Took the stage name Bowie—after the hero of the Alamo—to avoid being confused with Davy Jones of the Monkees. Recorded a version of "Little Drummer Boy" with Bing Crosby for Christmas 1977.*

> "Questioning my spiritual life has always been germane to what I was writing. Always. It's because I'm not quite an atheist and it worries me. There's that little bit that holds on: 'Well, I'm almost an atheist. Give me a couple of months.'"

Lieutenant-General William Boykin, *soldier of (1) Christ and (2) the U.S. Army. Born-again Christian. Appointed deputy undersecretary of defense for intelligence in June 2003.*

> "Why is this man [George W. Bush] in the White House? The majority of Americans did not vote for him. . . . I tell you this morning that he's in the White House because God put him there." *And we thought it was Katherine Harris, Karl Rove, James Baker, William Rehnquist & Co.!*

On Somali rebel leader Osman Atto, a Muslim:

"I knew that my God was bigger than his. I knew that my God was a real God and his was an idol." *I say, put both gods into the ring and let 'em slug it out.*

T. (Thomas) Coraghessan Boyle (1948–), *American novelist and short-story writer. Adopted the name Coraghessan when he was 17. No one knows why.*

"I am an atheist and a nihilist. . . . All my life I've been struggling to recover my faith or belief in something, but I can't. I wish I could, but I just can't. [*He can't. Leave him alone.*] I know KIERKEGAARD says you have to make the leap of faith, even though it's absurd. . . . But I believe in nothing. And it causes me tremendous despair and heartbreak. . . . There is nothing between us and the naked howling face of the universe. Nothing."

G. Richard Bozarth (1949–), *frequent contributor to MADALYN MURRAY O'HAIR's American Atheist magazine. He and his wife were married by O'Hair, in her (tax-exempting) capacity as a Universal Life minister, at the 1979 American Atheist Convention. (As it is fitting and proper. There's much too much intermarriage going on.)*

"Christianity has fought, still fights, and will fight science to the desperate end over evolution, because evolution destroys utterly and finally the very reason Jesus' earthly life was supposedly made necessary. Destroy Adam and Eve and the original sin, and in the rubble you will find the sorry remains of the son of god."

Charles Bradlaugh (1833–1891), *the most famous British atheist of his day. Cofounded the National Secular Society. Edited the secularist* National Reformer, *which the government unsuccessfully prosecuted for blasphemy and sedition. He and ANNIE BESANT were tried and sentenced to six months' imprisonment for publishing a birth control pamphlet; overturned on appeal, ending Britain's ban on disseminating contraceptive*

advice. Campaigned successfully to permit testifiers in court to "affirm" instead of taking the religious oath. Upon election to Parliament, refused to take the religious oath, therefore was prevented from taking his seat—despite winning four successive by-elections to fill it. Repeatedly arrested and/or escorted from the House by police officers. Last person in history to be imprisoned in Westminster's Clock Tower, a.k.a. Big Ben.

> "Atheism does not assert *no* God. The atheist . . . says 'I know not what you mean by God.'. . . The Bible God I deny; the Christian God I disbelieve in; but I am not rash enough to say there is no God as long as you tell me you are unprepared to define God to me."

> "I cannot follow you Christians; for you try to crawl through your life upon your knees, while I stride through mine on my feet."

Describing THOMAS HUXLEY's *new coinage, "agnosticism":*

> "A mere society form of Atheism."

Billy Bragg (1957–), *British folk/punk-rock musician of a distinctly red hue, and I don't mean Bush/Republican. Chosen by Woody Guthrie's daughter to record Guthrie's unfinished songs.*

> "The Soviet Union had many faults, and one was that it denied that people had spiritual needs. I went to Soviet museums of theology where they showed you how base religion was, while making a religion out of Lenin and Marx and never seeing the irony of that. . . . You don't want to live in a society based purely on materialism. Or, frankly, a society based purely on theology. . . . The United States looks like turning into both at once."

> "Our relationship to the Spirit [is] not a Him and Us thing. The Him is Us, probably. I can go both ways on whether there is a God. I say my prayers." *Also see* JOE HILL.

Nathaniel Branden (born Nathan Blumenthal, 1930–), *American psychologist. Leading promoter of the foul Objectivist philosophy of* AYN RAND, *with whom he had an affair. (He was 25 years her junior;*

both were married.) She designated him as her "intellectual heir," then expelled him from the movement in 1968 when she discovered he was cheating on her instead of just on his wife.

"Anyone who engages in the practice of psychotherapy confronts every day the devastation wrought by the teachings of religion."

Berkeley Breathed *(rhymes with "method"; 1957–), American cartoonist. Pulitzer–Prize-winning creator of the* Bloom County *and* Outland *comic strips. Described his politics as "shmiberal."*

"Although I'm an atheist, I don't fear death more than, say, sharing a room in a detox center with a sobbing Rush Limbaugh."

André Breton *(1896–1966), French surrealist writer, poet, and theorist. Author of* Surrealist Manifesto *(1924) which (belaboring the obvious) defined surrealism as "pure psychic automatism."*

"I have always wagered against God and . . . I am conscious of having won to the full. Everything that is doddering, squint-eyed, vile, polluted and grotesque is summoned up for me in that one word: God!"

Jacob Bronowski *(1908–1974), Polish-Jewish-British mathematician. Author and host of a popular book and TV series on the history of science,* The Ascent of Man, *which inspired* CARL SAGAN's Cosmos *series. Turned to biology to better understand the nature of violence after serving as an official observer of the after effects of the Nagasaki and Hiroshima bombings.*

"There is no absolute knowledge. And those who claim it, whether they are scientists or dogmatists, open the door to tragedy."

David Brooks *(1902–1994): Not, God forbid, the conservative* New York Times *columnist who wrote, "Religious groups should be*

sending out researchers to try to understand why there are pockets of people in the world who do not feel the constant presence of God in their lives, who do not fill their days with rituals and prayers"—not him, but the author of The Necessity of Atheism *(1933):*

> "By predicating a First Cause [God, the uncaused Cause], the theist removes the mystery a stage further back. . . . Such a belief is a logical absurdity, and is an example of the ancient custom of creating a mystery to explain a mystery. . . . Moreover, if it is reasonable to assume a First Cause as having always existed, why is it unreasonable to assume that the materials of the universe always existed? To explain the unknown by the known is a logical procedure; to explain the known by the unknown is a form of theological lunacy."

Andrew Brown, *English journalist and author. Winner of the first annual Templeton prize as best religious correspondent in Europe, 1995. Author of* The Darwin Wars. *Legally licensed to conduct baptisms, funerals, and weddings in the state of California by virtue of the five-minute process of getting ordained online, free, by the Universal Life Church of Modesto.*

> "My ordination material explained that 'Ministers are entitled to many discounts from retail agencies and various other trade entities and services. Among these are discounts on buses, trains, air travel, department store discounts, food discounts, retail and restaurant chains [and amusement parks]. . . . Ask and they may grant it.'"

Yasmin Alibhai-Brown (1949–), *Ugandan-born, London-based columnist of Ismaili Shi'ite Muslim background. Attacked both by Muslim groups and by right-wingers like columnist Michael Wharton, who said of his "invention," the "prejudometer," which detects and measures alleged racism: "At 3.6 degrees on the Alibhai-Brown scale, it sets off a shrill scream that will not stop until you've pulled yourself together with a well-chosen anti-racist slogan."*

> "By and large the lowest achieving community in this country are Muslims. When you talk to [Muslims] about why this is happening, the one reason they give you, the only reason, is Islamophobia.

Uh-uh. It is not Islamophobia that makes parents take 14-year-old bright girls out of school to marry illiterate men. . . ."

Robert Browning (1812–1889), *British poet and playwright. Fluent in French, Greek, Italian, and Latin by age 14. Became an atheist and a vegetarian (nonbeliever in the existence of meat) in imitation of his hero, PERCY BYSSHE SHELLEY. Brazenly declared, "I am no Christian"!!*

"Who knows most, doubts most."

"There's a new tribunal now / Higher than God's—the educated man's!"

Lenny Bruce (born Leonard Schneider, 1925–1966), *American comedian/satirist. His first of many arrests (generally for obscenity) was for impersonating a priest while soliciting donations for a leper colony (which actually existed) in British Guyana. Made around $8,000 in three weeks, sending $2,500 to the leper colony and keeping the rest: probably a more honest ratio than most religious charities.*

"Every day, people are straying away from the church and going back to God."

"If Jesus had been killed twenty years ago, Catholic school children would be wearing little electric chairs around their necks instead of crosses."

Giordano Bruno (1548–1600), *Italian philosopher, priest, astronomer, occultist. Revered by the irreverent as a freethought martyr. Argued that the stars are just like our Sun; that there are many worlds inhabited by intelligent beings; and that both space and time are infinite, leaving no room for Creation or Last Judgement. Burned at the stake by the Inquisition as a heretic. At his execution, wrote one biographer, "He turned his face away from the proffered crucifix and died in silence."*

"Perchance you who pronounce my sentence are in greater fear than I who receive it."

Around 420 b.e. (before Einstein):

> "There is no absolute up or down . . . no absolute position in space; but the position of a body is relative to that of other bodies. . . . there is incessant relative change in position throughout the universe, and the observer is always at the center of things."

William Jennings Bryan (1860–1925), *American lawyer, three-time Democratic Party nominee for president, Progressive Movement leader, champion of the little guy, scourge of big business, peace advocate . . . and fundamentalist ignoramus who prosecuted John Scopes for teaching evolution (see* CLARENCE DARROW) *and said things like:*

> "If the Bible had said that Jonah swallowed the whale, I would believe it."

> "If we have to give up either religion or education, we should give up education."

> "All the ills from which America suffers can be traced to the teaching of evolution." *Especially poverty and racial discrimination.*

Martin Buber (1878–1965), *influential Austrian-Jewish existentialist philosopher and scholar of Hasidic Judaism. Held that we generally relate to others and the world in "I-it" mode; the much rarer "I-thou" or "I-you" relationship is in effect an interaction with God, the eternal "you." (Not, as you might think, the eternal "who?")*

> "I don't like religion much, and am glad that in the Bible the word is not to be found."

> "The atheist staring from his attic window is often nearer to God than the believer caught up in his own false image of God." *My question is, why are atheists forced to live in attics?*

> "Since the primary motive of the evil is disguise, one of the places evil people are most likely to be found is within the church. What better way to conceal one's evil from oneself, as well as from others, than to be a deacon or some other highly visible form of Christian within our culture?"

John Buchan (1875–1940), *Canadian-born British government administrator in South Africa, member of Parliament, governor general of Canada, and a popular novelist whose most famous book,* The Thirty-Nine Steps, *was filmed by* HITCHCOCK.

"An Atheist is a man who has no invisible means of support."

James Buchanan (1791–1868), *15th U.S. president (1857–1861). Considered by many historians the worst president before George W. Bush, for letting the nation slide into civil war.*

"I have seldom met an intelligent person whose views were not narrowed and distorted by religion."

Robert William Buchanan (1841–1901), *Scottish-British poet, novelist, and playwright.*

"The gods are dead, but in their name / Humanity is sold in shame. / While (then as now!) the tinsel'd Priest / Sitteth with robbers at the feast, / Blesses the laden blood-stain'd board, / Weaves garlands round the butcher's sword, / And poureth freely (now as then) / The sacramental blood of men."

Pearl S. Buck (1892–1973), *American author. Brought up in China where her parents were Presbyterian missionaries. First American woman to win the Nobel Prize for Literature. Adopted seven children and established the first adoption agency dedicated to the placement of bi-racial children. A non-faith-based initiative.*

"I feel no need for any other faith than my faith in human beings."

"We send missionaries to China so the Chinese can get to heaven, but we won't let them into our country."

"Be born anywhere, little embryo novelist, but do not be born under the shadow of a great creed, not under the burden of original sin, not under the doom of Salvation."

Henry Thomas Buckle (1821–1862), *British historian. Author of* History of Civilization, *which he worked on ten hours a day for 17 years. So when he tells you the progress of civilization is directly proportional to the level of skepticism and inversely proportional to "credulity," you may depend upon it. One of the world's best chess players before he was 20.*

> "The clergy . . . have in all modern countries been the avowed enemies of the diffusion of knowledge, the danger of which to their own profession they, by a certain instinct, seem always to have perceived."

Buddha *(born Siddhartha Gautama* c. 623-– 543 B.C.E.), *Indian spiritual teacher. Following a midlife crisis, became Buddha at age 35, having already accumulated enough dharma and karma points (and having sat under a tree and vowed not to arise until he had found the Truth). May have been an avatar or incarnation of Lord Vishnu. But without DNA testing. . . .*

> "Do not believe in anything simply because you have heard it [or because it has] been handed down for many generations [or] because it is found written in your religious books [or] merely on the authority of your teachers and elders. But after observation and analysis, when you find that anything agrees with reason, and is conducive to the good and benefit of one and all, accept it and live up to it."

> "Doubt everything. Find your own light."

Nikolai Ivanovich Bukharin (1888–1938), *Russian Bolshevik revolutionary, Politburo member,* Pravda *editor; author of the textbook* The ABC of Communism, *still in use in most American elementary schools. Executed during Stalin's purges.*

> "Science has shown that religion began with the worship of dead ancestors. . . . The worship of dead rich men is thus the basis of religion."

Charles Bukowski (1920–1994), *American poet, novelist, and legendary boozer. At age 49, quit his job as a post office clerk in Los Angeles to write full time. Finished his first novel—titled* Post Office—*a month later. Eventually wrote more than 50 books. His funeral rites were conducted by Buddhist monks.*

> "I am my own God. We are here to unlearn the teachings of the church, state and our education system. We are here to drink beer. We are here to kill war. We are here to laugh at the odds and live our lives so well that Death will tremble to take us." *It's one of the Commandments. Thou shalt drink beer.*

Thomas E. Bullard (1949–), *American folklorist and collector of UFO abduction stories.*

> "Abduction reports sound like rewrites of older supernatural encounter traditions with aliens serving the functional roles of divine beings. Science may have evicted ghosts and witches from our beliefs, but it has just as quickly filled the vacancy with aliens having the same functions—it is business as usual in the legend realm where things go bump in the night."

Luis Buñuel (1900–1983), *Spanish-born filmmaker. Had a strict Jesuit education. It didn't take. His films are full of insults and injuries to priests, nuns, saints, why, to piety itself. Liked to walk around Paris dressed as a nun.*

> "God and Country are an unbeatable team; they break all records for oppression and bloodshed."

Near the end of his life:

> "Thank god I'm still an atheist."

Luther Burbank (1849–1926), *American horticulturist. Improved the world's food supply by developing new varieties of grains, fruits, vegetables—when he could have been spreading the word of God. Kept his views on religion to himself until the 1925 Scopes trial (see* CLARENCE DARROW), *then declared: "I am an infidel."*

"Mr. [WILLIAM JENNINGS] BRYAN [*who prosecuted John Scopes for teaching evolution*] was an honored friend of mine, yet this need not prevent the observation that the skull with which nature endowed him visibly approached the Neanderthal type. . . . Those who would legislate against the teaching of evolution should also legislate against gravity, electricity and the unreasonable velocity of light, and also should introduce a clause to prevent the use of the telescope, the microscope and the spectroscope. . . ."

Anthony Burgess (1917–1993), *British novelist, poet, screenwriter, critic, essayist, translator, composer, librettist, prodigious drinker and smoker. His best-known novel,* A Clockwork Orange, *was inspired by an incident during World War II when his wife was assaulted and robbed in London by U.S. Army deserters. Knew eight languages and some of five more. Known in Argentina as the British BORGES, who was known in Britain as the Argentine Burgess. (The two names are actually cognates.) When the two polyglots (believers in more than one glot) met, the story goes, rather than converse in English or Spanish, they settled on Old Norse. His writing method: "I start at the beginning, go to the end, then stop."*

"A perverse nature can be stimulated by anything. Any book can be used as a pornographic instrument, even a great work of literature. . . . I once found a small boy masturbating in the presence of the Victorian steel-engraving in a family Bible."

On the Vatican:

"All human life is here, but the Holy Ghost seems to be somewhere else."

Robert Burns (1759–1796), *the national poet of Scotland. Freemason. Inspired the founders of both liberalism and socialism. Alienated many of his friends by sympathizing with the French Revolution. Found a rhyme for purple. (I'm not telling.)*

"All religions are auld wives' fables."

"Why has a religioso [*sic*] turn of mind always a tendency to narrow and illiberalize the heart?"

John Burroughs (1837–1921), *American naturalist, writer, and poet. Native of New York's Catskill Mountains, where a range is named after him. Friend and biographer of WALT WHITMAN. Has been labeled a scientific pantheist (see ERNST HAECKEL).*

"Science has done more for the development of western civilization in one hundred years than Christianity did in eighteen hundred years."

"It is always easier to believe than to deny. Our minds are naturally affirmative."

"When I look up at the starry heavens at night . . . it is not the works of some God that I see there. I am face to face with a power that baffles speech. I see no lineaments of personality. . . . The universe is so unhuman . . . it goes its way with so little thought of man. . . . We must adjust our notions to the discovery that things are not shaped to him, but that he is shaped to them. The air was not made for his lungs, but he has lungs because there is air; the light was not created for his eye, but he has eyes because there is light. All the forces of nature are going their own way; man avails himself of them, or catches a ride as best he can. . . . We must and will get used to the chill . . . the cosmic chill. . . . Our religious instincts will be all the hardier for it."

"Every day is a Sabbath to me. All pure water is holy water, and this earth is a celestial abode."

William S. Burroughs (1914–1997), *American novelist/essayist/social critic. His cadaverous presence and wry, largely autobiographical narratives of drug addiction, homosexuality, and extreme alienation inspired/haunted two or three generations of outsiders and hipsters, from Beats to postmoderns and cyberpunks. "The only American novelist living today who may conceivably be possessed by genius," said NORMAN MAILER. Collaborated with Alan Ginsburg, Jack Kerouac, Laurie Anderson, John Giorno, Keith Haring, Kurt Cobain, Sonic Youth, and R.E.M. Shared an apartment with Kerouac in the 1940s, after his discharge from the Army for mental instability. Junkie for the last 40 years of his life. Shot and killed his common-law wife in a drunken game of William Tell in 1951. His novel*

Naked Lunch *(1959) features a stainless steel dildo called the Steely Dan.*
A Scientologist during the 1970s.

> "Son, never listen to a priest or a policeman . . . the only thing
> they have is the key to the shithouse."

> "If you're doing business with a religious son-of-a-bitch, get it in
> writing. His word isn't worth shit. Not with the good lord telling
> him how to fuck you on the deal."

> "The mark of a basic shit is that he has to be *right*."

> *In a good way?* "The perpetrating of miracles constitutes a brazen
> attempt to loosen up the universe. When you set up something as
> *miracle*, you deny the very concept of *fact*, establish a shadowy and
> spurious court . . . *beyond* the Court of Fact."

*Quoting a favorite figure, Hassan i Sabah, eleventh-century Persian religious
leader/warlord, founder of the Hashashin sect, which used hashish in the ini-
tiation of killers and gave us the word "assassin":*

> "Nothing is true, everything is permitted."

Sir Richard Burton (1821–1890), *British explorer, ethnologist, and*
*writer. Said to have spoken 29 European, Asian, and African languages. Trans-
lated the* Arabian Nights *and the* Kama Sutra *into English. As a soldier
serving in India, his expertise in local cultures and languages earned him the title
"the White Nigger" among fellow soldiers. Made the Hajj to Mecca disguised in
Eastern dress; even had himself circumcised to reduce the risk of, well, exposure.*

> "The more I study religions, the more I am convinced that man
> never worshipped anything but himself."

Samuel Butler (1835–1902), *English novelist, author of* Erewhon
and The Way of All Flesh. *Was preparing for ordination to the Anglican
clergy when "he discovered that baptism made no apparent difference to the
morals and behaviour of his peers and began questioning his faith."*[1]

> *Reversing Alexander Pope:* "An honest God's the noblest work
> of man."

"If God wants us to do a thing, he should make his wishes sufficiently clear. Sensible people will wait till he has done this before paying much attention to him."

"Prayers are to men as dolls are to children. They are not without use and comfort, but it is not easy to take them seriously."

"Belief like any other moving body follows the path of least resistance."

"What is faith but a kind of betting or speculation after all? It should be, 'I bet that my Redeemer liveth.'"

"People are equally horrified at hearing the Christian religion doubted, and at seeing it practiced."

"Christ: I dislike him very much; still I can stand him. What I cannot stand is the wretched band of people whose profession is to hoodwink us about him."

"Christ and The Church: If he were to apply for a divorce on the grounds of cruelty, adultery and desertion, he would probably get one."

"I really do not see much use in exalting the humble and meek; they do not remain humble and meek long when they are exalted."

"Christ was only crucified once, and for a few hours. Think of the thousands he has been crucifying in a quiet way ever since."

"He has spent his life best who has enjoyed it most. God will take care that we do not enjoy it any more than is good for us."

"There is nothing which at once affects a man so much and so little as his own death."

"Death is only a larger kind of going abroad."

"Theist and Atheist. The fight between them is as to whether God shall be called God or shall have some other name."

Lord Byron *(born George Noel Gordon, 1788–1824), English Romantic poet and adventurer. Kept a bear while a student at Trinity College, Cambridge, in compliance with the rule forbidding pet dogs.*

"Of religion I know nothing—at least, in its favor."

"All are inclined to believe what they covet, from a lottery ticket up to a passport to Paradise."

A religious zealot in his poem Childe Harold: "I hope to merit Heaven by making earth a Hell."

"The basis of your religion is injustice. The Son of God the pure, the immaculate, the innocent, is sacrificed for the guilty. This proves his heroism, but no more does away with man's sin than a school boy's volunteering to be flogged for another would exculpate a dunce from negligence."

On those to be damned at the Last Judgment: "When the World is at an end, what moral or warning purpose can eternal tortures answer?"

NONBELIEVER * HUMANIST * RATIONALIST * FREETHINKER * AGNOSTIC * GODLESS * HERETIC * ATHEIST * SECULAR * INFIDEL *

C

Herb Caen (1916–1997), *American journalist/columnist. Spent most of his career with the* San Francisco Chronicle. *Pulitzer Prize, 1996. Credited with inventing the words "beatnik" and "hippie."*

"The trouble with born-again Christians is that they are an even bigger pain the second time around."

L. Sprague de Camp (1907–2000), *American science fiction and fantasy author. Wrote over 100 novels as well as nonfiction on topics such as racism. (Pointed out that no scholar had ever sought to prove that his own ethnicity was inferior to others.) Enjoyed debunking claims of the supernatural.*

"It does not pay a prophet to be too specific."

Joseph Campbell (1904–1987), *American scholar of mythology. Believed all religions and deities are "masks" of the same transcendent, "unknowable" truth, and have a common ancestry. Read for nine hours a day for five straight years during the Great Depression, which in fact was named for the state of mind he thus achieved. His book* The Hero with a Thousand Faces *was inspired by* JAMES JOYCE'S Finnegans Wake *and in turn*

inspired the insufferable cartoon mythology of George Lucas' Star Wars. (See BBC ONLINE.)

"Mythology is what we call someone else's religion."

"Religion can be defined as misinterpreted mythology. . . . [It attributes] historical references to symbols which properly are spiritual in their reference."

Albert Camus (1913–1960), *French existentialist author. Preached a heroic atheism. In his story "The Myth of Sisyphus," the hero, who is condemned to spend eternity trying to push a boulder up a mountain, fully conscious that he has no hope of success or of a decent wage, represents man in a godless world.*

But he is scornful of the gods, his tormentors. No sissyphus, this Sisy: "The lucidity that was to constitute his torture at the same time crowns his victory. . . . Crushing truths perish from being acknowledged. . . . All Sisyphus' silent joy is contained therein. His fate belongs to him. . . . The struggle itself toward the heights is enough to fill a man's heart. One must imagine Sisyphus happy."

"There is only one religion that exists throughout all history, the belief in eternity. This belief is a deception."

"If there is a sin against life, it consists perhaps not so much in despairing of life as in hoping for another life and in eluding the implacable grandeur of this life."

"Don't wait for the Last Judgement. It takes place every day."

Carlo Cardinal Caraffa (1519–1561), *Roman Catholic cardinal; nephew (Gr. nepos > nepotism) of Pope Paul IV.*

"Populus vult decipi, decipiatur." "The people want to be deceived. Let them be deceived." *(Often misattributed to Karl Rove.)*

Asia Carerra (born Jessica Andrea Steinhauser, 1973–), *retired "adult" film actress with 275 films, uh, under her belt. Most recommended:*

Whoriental Sex Academy 6 and Bangkok Booberella. *Performed as a pianist at Carnegie Hall at age 13. Ran away from home at 17. Member of Mensa. Really. (Membership requires a minimum score of 38C. No, sorry, that's Immensa.)*

> "I've always been an atheist. Science explains everything. There is no meaning in life except to be the best at something. If only I could be the best at something, perhaps my parents would love me."

> "People around the world are extremely gullible . . . I need to get off my ass [*or did she say "back"*] and start a religion of my own!"

George Carlin (1937–), *American comedian. First-ever host of* Saturday Night Live *(1975). Author of* When Will Jesus Bring the Pork Chops? *(2004).*

> "When it comes to believing in God, I really, really tried . . . but . . . the more you look around, the more you realize . . . something is wrong here. War, disease, death, destruction, hunger, filth, poverty, torture, crime, corruption, the Ice Capades. . . . This is not good work. If this is the best God can do, I am not impressed. Results like these do not belong on the résumé of a Supreme Being."

> "Atheism is a non-prophet organization."

> "You know who I pray to? Joe Pesci. . . . He looks like a guy who can get things done. Joe Pesci doesn't fuck around. . . . For years I asked God to do something about my noisy neighbor with the barking dog. Joe Pesci straightened that cocksucker out with one visit."

Thomas Carlyle (1795–1881), *Scottish essayist and historian. His lament that society was being dehumanized and communal values were being destroyed by impersonal economic forces and ruthless capitalism, and his distaste for democracy and belief in charismatic leadership, influenced socialism, conservatism, and fascism—including Hitler himself. The completed manuscript for his two-volume* French Revolution, A History *was accidentally*

burned by JOHN STUART MILL'*s maid; Carlyle rewrote it from scratch.
All this from a non-atheist, or non-non-theist:*

"God does nothing." *(Doesn't even protect manuscripts . . .)*

"Just in the ratio that knowledge increases, faith diminishes."

"If Jesus Christ were to come today, people would not even
crucify him. They would ask him to dinner, and hear what he had
to say, and make fun of it."

"If I had my way, the world would hear a pretty stern
command—Exit Christ."

"There is one true church of which at present I am the only
member."

Andrew Carnegie (1835–1919), *Scottish-American steel mag-*
nate and philanthropist. Professed atheist.

"I have not bothered Providence with my petitions for about
forty years."

"I give money for church organs in the hope that the organ
music will distract the congregation's attention from the rest of
the service."

Adam Carolla (1964–), *American comedian. Former cohost of the*
radio show Loveline *and* The Man Show, *the TV show for unregenerate*
male chauvinist swine; host of the Adama Carolla *morning radio show,*
which replaced HOWARD STERN *on some stations. Graduated high school*
(where he majored in ceramics) with a 1.75 GPA.

"Nah, there's no bigger atheist than me. Well, I take that back.
I'm a cancer screening away from going agnostic and a biopsy
away from full-fledged Christian." *(No atheist in oncology clinics?
See* LANCE ARMSTRONG, MELISSA ETHERIDGE, RUTH
HURMENCE GREEN*. . .)*

"I'm very insulted when people say, 'Well, without religion
what's to stop people?'. . . . How many people on death row are
atheists? They all love Jesus and they all put a shiv into a Korean

liquor store owner. . . . I'll tell you where God is. God is in jail! Someone does time, then all of a sudden he's a Born-Again Christian!" *(See W. T. Root.)*

Angela Carter (1940–1992), *English journalist, and postfeminist magical realist novelist.*

"Mother goddesses are just as silly a notion as father gods. If a revival of the myths of these cults gives women emotional satisfaction, it does so at the price of obscuring the real conditions of life. This is why they were invented in the first place." *Is that decoded clearly enough for you, Dan Brown? (And you, Riane Eisler and Robert Graves?)*

Dick Cavett (1936–), *American talk show host. Grandson of a fundamentalist Baptist preacher. His most famous stand-up line: "I went to a Chinese-German restaurant. The food is great, but an hour later you're hungry for power."*

"It would be wonderful to believe. . . . It would make life easier, it would explain everything, it would give meaning. . . . But something about knowing it could instantly make me much happier makes it somehow unworthy of having."

Charlie Chaplin (1889–1977), *British-born American film actor, producer, and director. At age 28, married a 16-year-old. At 35, married another 16-year-old. (Yes, he'd divorced the first one. What kind of tramp do you think he was?) Hitler watched Chaplin's satire* The Great Dictator *(1940) twice. Three months after his death, his body was stolen in a failed attempt to extort money from his family.*

"By simple common sense, I don't believe in God."

"I would love to play the part of Jesus! I fit it perfectly because I am a comedian." *(And Jewish, and left wing.)*

Noam Chomsky (1928–), *American linguist, MIT professor, and left-wing political activist. The most important theoretical linguist of the past century. Grew up immersed in Hebrew culture and literature. An anarcho-syndicalist from the age of 12 or 13—when most boys are still Spartacist-Trotskyist. Especially critical of U.S. foreign policy.*

"Do I believe in God? . . . I don't understand the question."

"How do I define God? I don't. . . . I see no need. . . . As for 'First Principles,' basing them on divinities is, I think, a very bad idea. That leaves anyone free to pick the 'first principles' they choose on other grounds, and to disguise the choices as 'what God commands.' . . . Nothing is gained . . . and a great deal is lost: specifically, the opportunity to question, elaborate, modify, or reject them. . . . If you want to use the word 'God' to refer to 'what you are and what you want'—well, that's a terminological decision, not a substantive one."

"I would agree with the classic anarchist slogan 'Ni dieu, ni maitre' [No god, no master]."

"That 'religion is inherently irrational' is surely true. Why one set of beliefs that are offered without argument or evidence rather than another?"

"The Bible is basically polytheistic, with the warrior God demanding of his chosen people that they not worship the other Gods and destroy those who do. . . . It would be hard to find a more genocidal text in the literary canon. . . .'" *Also see epigraph at the top of this volume's Introduction.*

Jesus Christ (c. 4 B.C.–33 A.D.), *Jewish religious teacher, reformer, reputed miracle worker and Messiah who founded a new sect of Judaism.*

"Think not that I am come to send peace on earth; I came not to send peace, but a sword."—*Matthew 10:34.*

Sir Winston Churchill (1874–1965), *British Conservative (mostly) politician and historian. Prime minister, 1940–1945, 1951–1955. Tried to convince President Truman to drop atomic bombs on major Soviet cities after WWII. Voted Greatest Briton ever in a 2002 BBC poll. Described on the Churchill Centre Web site as "having made so many deposits in the bank of Religion" as a youth that he had been "confidently withdrawing from it ever since, never bothering to check the balance—there might indeed be an overdraft"—which could have meant a fine and suspension of check-writing privileges. Called himself an "optimistic agnostic" and "not a pillar of the Church but more of a flying buttress—I support it from the outside."*

> "Man will occasionally stumble over the truth, but most of the time he will pick himself up and continue on."

Upon meeting with Saudi King Ibn Saud and being told that the king would not allow strong drink:

> "[I understand but] my religion prescribed as an absolute sacred rite smoking cigars and drinking alcohol before, after, and if need be during all meals and the intervals between them."

Marcus Tullius Cicero (106–43 B.C.E.), *Roman statesman/ philosopher/lawyer; considered the greatest Latin orator and prose stylist. His political enemy Marc Antony had him executed. His head and hands were displayed in the Forum. They say Antony's wife pulled out his tongue and jabbed it repeatedly with her hairpin, taking a final revenge against his oratical prowess. Roman women, man.*

> "Without the hope of immortality, no one would ever face death for his country." *(Or his Allah.)*

> "I wonder that a soothsayer doesn't laugh whenever he sees another soothsayer."

> "When we call corn Ceres or wine Bacchus, we use a common figure of speech, but do you imagine that anyone is so mad as to believe the thing he feeds upon is a god?" *Or that what he drinks (at communion) is the god's blood?*

Emil Cioran (1911–1995), *Romanian-born French philosopher, writer, essayist. Son of an Orthodox priest. Flirted with Romania's far-right Iron Guard in the 1930s and early 1940s, but never actually asked it out on a date. Joined a philosophical movement that fused existentialism with elements of fascism. His mother reportedly told him if she had known he was going to be so unhappy she would have aborted him. This "made an extraordinary impression which led to an insight about the nature of existence,"[1] and to his statement,*

"I'm simply an accident. Why take it all so seriously?"

"Bach's music is the only argument proving the creation of the Universe can not be regarded a complete failure. . . . Without Bach, God would be a second-rate figure."

"The fact that life has no meaning is a reason to live—moreover, the only one."

"Life inspires more dread than death—it is life which is the great unknown."

"What does the future, that half of time, matter to the man who is infatuated with eternity?"

"The fear of being deceived is the vulgar version of the quest for Truth." *(He's talking about us atheists among others, isn't he?)*

"No human beings are more dangerous than those who have suffered for a belief: the great persecutors are recruited from the martyrs not quite beheaded."

Arthur C. Clarke (1917–), *British science-fiction author and inventor. Ardent atheist. The film* 2001: A Space Odyssey *was based on his short story "The Sentinel." In the future envisioned in his novel* 3001, *religion has become taboo; the blood-soaked religions of the past, are viewed as barbaric. The dinosaur species* Serendipaceratops arthurcclarkei *is named after him, or else it's an extraordinarily serendipitous coincidence.*

"I don't believe in God but I'm very interested in her."

"I suspect that religion was some random by-product of mammalian reproduction . . . a necessary evil in the childhood of our species. . . . but why was it more evil than necessary? Isn't killing people in the name of God a pretty good definition of insanity?"

"I would defend the liberty of consenting adult creationists to practice whatever intellectual perversions they like in the privacy of their own homes; but it is also necessary to protect the young and innocent."

Noting that in Sri Lanka, where he lives, "everyone" believes in reincarnation:

"The problem with reincarnation is that it's hard to imagine what the storage medium for past lives would be. Not to mention the input-output device." *You're an inventor? Invent one, damn it!*

Jeremy Clarkson (1960–), *British automotive writer/critic. Host of the BBC car show* Top Gear, *which has 350 million viewers worldwide and won an International Emmy in 2005. Described by a U.K. newspaper as a "dazzling hero of political incorrectness." Known for publicly smoking as much as possible on National No Smoking Day. Boasted on TV of eating whale topped with grated puffin, and seal flipper, which he said tastes "exactly like licking a hot Turkish urinal." (Turning eco-green yet?) Installed a vintage fighter jet on his front lawn; tried telling the irate town council it was a leaf blower. Has said Mark Chapman's hit list should have listed Paul McCartney above JOHN LENNON. Wrote in an editorial that "most Americans barely have the brains to walk on their back legs." His book* I Know You Got Soul *is about the soul he believes many machines have.*

"The church, to me, is just loathsome. I really do object to the fact that it's so powerful. Even though it actually represents fewer people than the local Tufty Club. And they quote from this book which is a fucking fairy tale. It's all just rubbish. You can't walk on water 'cause you'd sink. When you're dead, you can't come back to life. It's that simple."

Voltairine de Cleyre (1866–1912), *American women's rights and labor activist, renowned atheist lecturer, and "the most gifted and brilliant anarchist woman America ever produced," according to socialist/anarchist EMMA GOLDMAN, who wrote her biography. Her father named her after his favorite skeptic but, unable to support the family, sent her to a Catholic convent/boarding school, an experience that, naturally, pushed her toward atheism.*

"The question of souls is old—we demand our bodies, now. We are tired of promises, God is deaf, and his church is our worst enemy."

William Kingdon Clifford (1845–1879), *pioneering English mathematician and philosopher. Proposed a geometric, "curved space" theory of gravitation decades before Einstein. Argued that mind is the ultimate reality; an atom of matter is ultimately an atom of the same "mind-stuff" that thought and feeling are made of. Regarded by the clergy as a dangerous champion of antispiritual scientific tendencies.*

"If a man, holding a belief which he was taught in childhood or persuaded of afterwards, keeps down and pushes away any doubts which arise about it in his mind, purposely avoids the reading of books and the company of men that call in question or discuss it . . . the life of that man is one long sin against mankind."

Alexander Cockburn (1941–), *Scottish-born U.S.-based Irish left-wing journalist.* Nation *magazine columnist,* CounterPunch *coeditor. On Los Angeles school officials pulling an edition of the Koran from the district's libraries because of "complaints" that the footnotes are anti-Semitic (oh, surely not):*

"It surely won't be long before the Bible is pulled as well, since the Old Testament is rough on the Palestinians [*sic*] and the New Testament rough on the Jews. . . . Though my basic view is that any childish mind not innoculated [*sic*] by compulsory religion is open to any infection, by all means let us sweep the Bible and the Koran off every bookshelf whither might stray the hand of impressionable youth. Such a cleansing act would return us to the very roots of the European enlightenment."

Jean Cocteau (1889–1963), *French writer, dramatist, filmmaker, designer, and of course, boxing manager. (Really.) Openly gay, openly Surrealist.*

"Mystery has its own mysteries, and there are gods above gods. We have ours, they have theirs. That is what's known as infinity."

"Man seeks to escape himself in myth, and does so by any means at his disposal. Drugs, alcohol, or lies. Unable to withdraw into himself, he disguises himself. Lies and inaccuracy give him a few moments of comfort."

Andrei Codrescu (1946–), *Romanian-born American poet, commentator for National Public Radio, professor of English at Louisiana State University, and former editor of the online journal* The Exquisite Corpse.

"The evaporation of four million [people] who believe in this crap [Christian Rapture, or Crapture] would leave the world a better place."

Chapman Cohen (1868–1954), *president of the National Secular Society, Britain's largest atheist organization, 1915–1949. Editor of* Freethinker *magazine.*

"Gods are fragile things; they may be killed by a whiff of science or a dose of common sense."

"There are still many who continue to marvel at the wisdom of God in so planning the universe that big rivers run by great towns, and that death comes at the end of life instead of in the middle of it."

"Atheists decline to be as miserable as Christians assure them they ought to be. . . . Christians are not only perplexed at the sight of happy atheists, they are annoyed. If atheists are happy, it must be because they lack the fine moral development of the Christian. If they were only better, they should know how poor is the happiness they feel. . . . But the happiness of an atheist [expresses] a disposition that has ceased to torture itself with foolish fancies, or perplex itself with useless beliefs."

Edmund D. Cohen, *American psychologist, lawyer, and one-time fundamentalist Christian. Author of* The Mind of the Bible Believer *(1988), which displays this blurb by* FRANK ZAPPA: *"Edmund Cohen was the first, to my knowledge, to sort out the hints of ominous intentions in Pat Robertson's antics. . . . I expect Cohen to be the first to detect it when the next of these cockroach-messiahs is hatched and descends from the baseboards."*

"The content of the [fundamentalist] teaching, as well as the form of social relations, is set up so as to dig a psychological moat around the believers. . . . The supposed renewal of the mind so that it thinks only godly thoughts, the fatuous peace and tepid joy of the person exhibiting euphoric calm, the apparent absence of friction with other people, these are side effects of a dissociated state of mind."

"[If one is mindful of the biblical God's] vengefulness and his extravagant sadomasochism . . . one ends up straightjacketing one's thoughts and feelings lest something taboo be exhibited [to God's] unremitting scrutiny. At the seminary, the constant effort I expended to censor my thoughts and feelings began to exhaust me—a real self-induced neurosis. . . . Devoutness had brought out some of my worst tendencies. I had grown not in righteousness, but in rigidness, not in purity, but in priggish-ness—not in holiness, but in ass-holiness! I had been on my way to becoming a regular little Savonarola, a regular little Calvin itching to get a Servetus burned at the stake. . . . No wonder that the accomplishments of contemporary fundamentalists in the arts and sciences are so pathetically few . . ."

Larry Cohen (1938–), *American filmmaker. Gave the world* It's Alive; It Lives Again; It's Alive III: Island of the Alive; *and* God Told Me To, *in which murderers who believe God told them to kill but who are in fact possessed by a Demon from Outer Space commit a series of bizarre homicides.*

"Everybody's got a different idea what God thinks and what God likes and what God is; the crazy guy on the corner knows about as much as the guy in St. Patrick's Cathedral—none of them know anything. People say, 'Reverend Moon—what a crook!' and I say, 'But what about the Pope?' It's all the same. Anybody who starts telling you what God thinks should be locked up immediately." *Speaking of which: also wrote the classic* Women of San Quentin, *or as it might be called,* Les Ms.

Morris R. Cohen (1880–1947), *American professor of philosophy and of law at City College of New York (whose reputation as the "prole-tarian Harvard" he helped build) and the University of Chicago. Studied with* WILLIAM JAMES *at Harvard. Legendary for his ability to demolish philosophical systems—but also had a positive message.*

"There is obviously an important difference between an establishment [i.e., science] that is open . . . and one that regards the questioning of its credentials as due to wickedness of heart, such as [Cardinal] Newman attributed to those who questioned the infallibility of the Bible. Rational science treats its credit notes as always redeemable on demand, while non-rational authoritarianism regards the demand for the redemption of its paper as a disloyal lack of faith."

"If religion cannot restrain evil, it cannot claim effective power for good."

Stephen Colbert (1964–), *American comedian. Host of Comedy Central's faux-Faux News sendup, the* Colbert Report. *Former* Daily Show *"correspondent" known for his "This Week in God" sketch, featuring the all-knowing God Machine. His speech lampooning Bush and the press corps to their faces at the 2006 White House Correspondents' Dinner became the #1 download on iTunes. Named one of* Time *magazine's 100 most influential people for 2006. Practicing Roman Catholic. The Atheists' first draft pick for 2007. From his White House Correspondents' speech:*

"Though I am a committed Christian, I believe that everyone has the right to their own religion, be you Hindu, Jew or Muslim. I believe there are infinite paths to accepting Jesus Christ as your personal savior."

Samuel Taylor Coleridge (1772–1834), *English Romantic poet, critic, and philosopher. Wrote his famous poem* Kubla Khan *while high on opium, which he was . . . quite fond of. Considered a "giant among dwarfs" by his contemporaries.*

"Not one man in ten thousand has either strength of mind or goodness of heart to be an atheist."

"Our quaint metaphysical opinions, in an hour of anguish, are like playthings by the bedside of a child deathly sick."

Lucy Colman (1817–1906), *American suffragist and abolitionist.*
"Often mobbed, she found that the racist ringleaders were nearly always cler-gymen." [5] *Scandalized America by urging the abolition of corporal punish-ment in schools. Dated her articles in* The Truth Seeker, *the leading freethought publication, from the year of* GIORDANO BRUNO's *martyrdom (e.g., 1887 became 287).*

> "I had given up the church, more because of its complicity with slavery than from a full understanding of the foolishness of its creeds."

> "If your Bible is an argument for the degradation of woman, and the abuse by whipping of little children, I advise you to put it away, and use your common sense instead."

Charles Caleb Colton (1780–1832), *English cleric. Wrote phenom-enally successful books of essays and aphorisms. Made and lost fortunes gambling and dealing in art. Wound up broke. Took ill and committed suicide rather than undergo (preanesthetic) surgery. (God gave us pain. Science gave us anesthetics.)*

> "He that dies a martyr proves that he was not a knave, but by no means that he was not a fool."

> "Some reputed saints that have been canonized ought to have been cannonaded."

Cyril Connolly (1903–1974), *English writer and literary critic.*

> "Those of us who were brought up as Christians and who have lost our faith have retained the Christian sense of sin without the saving belief in redemption. This poisons our thought and so paralyzes us in action."

Joseph Conrad (born Józef Teodor Konrad Korzeniowski, *1857–1924) Polish-born British novelist. Parents were Polish nobility. Orphaned at age 11. Became a seaman at 17 to escape 25-year conscription into the Russian army. Became involved in gunrunning, political conspiracy, and (had he no scruples at all?) literature.*

"The belief in a supernatural source of evil is not necessary; men alone are quite capable of every wickedness."

"God is for men and religion for women."

Hans Conzelmann (1915–1989), *German New Testament scholar. Argued that Jesus did not regard himself as Messiah or Son of God.*

"The Christian community continues to exist because the conclusions of the critical study of the Bible are largely withheld from them."

Harvey Cox, Jr. (1929–), *leading American theologian. Professor of divinity, Harvard Divinity School.*

"Sermons remain one of the last forms of public discourse where it is culturally forbidden to talk back."

Michael Crichton (1942–), *American novelist, TV and film producer and director, and Harvard-trained physician. As an undergraduate, plagiarized a work by* GEORGE ORWELL *and submitted it to a professor he suspected was deliberately giving him low grades. It got a B-. Sullied his fine a-religious name in recent years by campaigning against environmentalism and denying global warming (thus endearing himself to George W. Bush, who invited him to the White House).*

To Paul Rifkin, author of The God Letters, *who, pretending to be a fifth grade pupil working on a school project, wrote to celebrities asking if they believed in God:*

"Organized religion is a business and nothing else, unless you want to think of it as a way to organize wars efficiently. Also, organized religion tells you what to think, and I believe the only way to know about God is to find out on your own. What somebody else tells you is of very little use. . . . I believe that we are all God, and God is all of us. But this means you can find out everything you need to know by yourself, by what you see and what you feel. Nobody knows any more than you do—even though you're just in the fifth grade"

Francis Crick, (1916–2004), *British physicist, molecular biologist, and neuroscientist. Nobel-winning co-discoverer of the structure of DNA. Professed "agnostic with a strong inclination toward atheism" since age 12. Has wondered whether a computer might be programmed so as to have a soul. Speculated that it might be possible to detect chemical changes in the brain of a person praying. Wrote of a possible new direction for research which he termed "biochemical theology." The field of "neurotheology" has since emerged. Advocated for the establishment of Darwin Day as a British national holiday. (As they have done in Alabama and Tennessee. Not!)*

> "If revealed religions have revealed anything it is that they are usually wrong."

Quentin Crisp (born Denis Charles Pratt, 1908–1999), *English writer, actor, and wit. His flamboyant and unabashedly effeminate homosexuality, as described in his 1968 autobiographical book* The Naked Civil Servant, *made him a gay icon. Never did any housework because, he said, the dirt didn't get any worse after the first four years. Habitually answered the phone with the phrase "Yes, God?"—"just in case," he explained. Played Elizabeth I in the 1992 film* Orlando. *Sting's song* Englishman in New York *is about him.*

> "When I told the people of Northern Ireland that I was an atheist, a woman in the audience stood up and said, 'Yes, but is it the God of the Catholics or the God of the Protestants in whom you don't believe?'"

Aleister Crowley (1875–1947), *English occultist (Kabbalah, "magick," astrology, you name it), writer, sexual revolutionary, cult leader, chess master, mountain climber, painter, heroin and opium addict, bisexual, and bigot. After the Bolsheviks took power in Russia, he wrote to* TROTSKY *offering to rid Russia of Christianity. The press dubbed him "The Wickedest Man In the World."*

> "If one were to take the Bible seriously one would go mad. But to take the Bible seriously, one must be already mad."

> "I slept with Faith, and found a corpse in my arms on awaking; I drank and danced all night with Doubt, and found her a virgin in the morning."

NONBELIEVER * HUMANIST * RATIONALIST * FREETHINKER * AGNOSTIC * GODLESS * HERETIC * ATHEIST * SECULAR * INFIDEL *

C. W. Dalton, *American atheist writer. Author of* The Right Brain and Religion *and* You're OK—The World's All Wrong*; contributor to* Truth Seeker *magazine, "world's oldest freethought publication."*

> "The life of an atheist seems especially tragic. . . . The atheist cannot even look forward to being vindicated. Even if he is right, religionists will die never finding out that they had worshiped a God that didn't exist, had prayed to the wind, had tithed to support priestly parasites. . . . No, atheists know they will never get even this satisfaction. [*No false advertising from this religion.*] Perhaps it is time for atheists to give up on eradicating religion and accept it as they accept other incurable diseases and as they accept bad weather, taxes and in-laws as inevitable."

Clarence Darrow (1857–1938), *American lawyer. Best known for defending teenage thrill killers Leopold and Loeb and for the only pro bono case he ever took, defending high school teacher John Scopes in the 1925 Monkey Trial, in which Scopes, as Darrow remarked, was on trial "for the crime of teaching the truth." Lost only one out of more than 100 murder cases.*

> "If today you can take a thing like evolution and make it a crime to teach in the public schools, tomorrow you can make it a crime to teach it in the private schools and next year . . . to teach it to the hustings or in the church. At the next session you may ban

books and the newspapers. . . . Ignorance and fanaticism are ever busy and need feeding."

Echoing HUME: "Since man ceased to worship openly an anthropomorphic God and [instead] talked vaguely and not intelligently about some force in the universe, higher than man, that is responsible for the existence of man and the universe, he cannot be said to believe in God."

"I don't believe in God because I don't believe in Mother Goose."

"I do not consider it an insult, but rather a compliment, to be called an agnostic. I do not pretend to know where many ignorant men are sure."

"Some of you say religion makes people happy. So does laughing gas. So does whiskey."

"Every man knows when his life began. . . . If I did not exist in the past, why should I, or could I, exist in the future [after death]?"

"When we accept the fact that all men and woman are approaching an inevitable doom [death], the consciousness of it should make us more kindly and considerate of each other."

Charles Darwin (1809–1882), *English naturalist who discovered the chief mechanism of evolution. Was studying for the clergy when he signed on as naturalist aboard the government research ship H.M.S. Beagle. Became disillusioned (literally) during the course of his circumnavigation of the globe (1831–1836), and by age 40 was, in his own words, "a complete disbeliever in Christianity" and a professed agnostic. "Darwin said, Yes, there's fantastic design in the biosphere, and I'm going to show you how you can get that design without a designer."*— DANIEL DENNETT. *In 2000, Darwin's image replaced* DICKENS'S *on the British ten-pound note—partly because Darwin's beard would be harder to forge. Overnight, thousands of Dickens beard forgers became unemployed.*

"My theology is a simple muddle. I cannot look at the universe as the result of blind chance, yet I can see no evidence of beneficent design, or indeed of design of any kind."

"There seems to me too much misery in the world. I cannot persuade myself that a beneficent and omnipotent God would have designedly created the ichneumonidae with the express intention of their feeding within the living bodies of caterpillars, or that a cat should play with a mouse."

"We can allow satellites, planets, suns, [the] universe, nay whole systems of universe[s], to be governed by laws, but the smallest insect, we wish to be created at once by special act."

On the mind arising from the brain, not the "soul":

"Why is thought being a secretion of brain, more wonderful [i.e., strange] than gravity a property of matter? It is our arrogance, our admiration of ourselves [that prevents our accepting it]."

Last words:

"I am not the least afraid to die."

King David (c. 1011–971 B.C.E.), *second king of the united kingdom of Israel. Traditionally regarded as the author of the Psalms (more likely written c. 600 B.C.E.).*

"For I have done your bidding, I have slain mine enemies in your name. I have put women and children to death in your honor, I have caused great pain among them, for your glory."—*Psalms, 5:4–10*

Robertson Davies (1913–1995), *Canada's leading man of letters over the latter half of the twentieth century. His greatest novel,* Fifth Business, *considers what might happen if a true saint appeared and tried to live in a small Canadian town in the early 1900s.*

"Fanaticism is overcompensation for doubt." *"All fanaticism is a strategy to prevent doubt from becoming conscious."—H. A. Williams, former dean of Trinity College, Cambridge.*

Walter A. Davis, *emeritus professor of English, Ohio State University. His 2006 book* Death's Dream Kingdom: The American Psyche

since 9-11 examined, in his words, "the apocalyptic scenario driving the Christian fundamentalist assault on our basic freedoms."

"The question that constitutes the inherent and lasting fascination of religion [is] *not what people believe, but why*. . . . Religion is invaluable because it offers the deepest insight into the nature of the psyche and its needs."

"One irony of fundamentalist reading is the rather considerable constraints it places on the deity. He proclaimeth and what He says remains so forever, beyond development, change, revision. Whatever abomination of sex hatred one unearths from Leviticus must remain gospel today. . . . After all, 'It's in the Bible.' That repeated assertion expresses the essence and fundamental paralysis of the literal mind."

"Fundamentalist certitude always becomes rectitude; and the Bible is mined for all the things one can label abomination. Thereby a sensibility that wants to have nothing to do with the world takes revenge upon it."

"The power of [evangelical Christian] conversion to produce a saved self makes the Catholic confessional [look like] the operation of rank amateurs. There . . . one gets temporary relief from sins . . . but not a lasting transformation."

"The obsessional need to preach . . . to let every stranger one meets know as soon as possible that one is a born-again Christian, are practices that derive not from a lack of social skills but from a manic necessity. . . . Without evangelical activity the fundamentalist psyche sinks into a state of empty boredom. . . . Thus the lassitude of Dubya before 9-11 and the hectic messianic energy that has defined him since. . . . God has chosen one not just to convert the World but to wage war on whatever one labels evil. . . . We should all indeed be trembling in our boots to know the mind-set that now has its finger on the nuclear trigger."

Richard Dawkins (1941–), *Kenyan-born British zoologist and evolutionary theorist. His landmark book* The Selfish Gene *argued that the gene, not the individual plant or animal, is the principal unit of selection in evolution and may be regarded in effect as the organism; we plants (and animals) as merely vehicles built by DNA molecules to transport them around and supply*

them with food, oxygen, and other DNA to "mate" with. (Or as Groucho Mark is said to have said: Life is the whim of several brillion cells to be with you for a while." Or as ANONYMOUS *said: " A chicken is an egg's way of producing more eggs.") Dawkin's friend* DOUGLAS ADAMS *described reading the book as "one of those absolutely shocking moments of revelation when you understand that the world is fundamentally different from what you thought it was . . . almost like a religious experience." Dawkins's book* The Blind Watchmaker *demolished the creationists' "argument from design." Dubbed "Darwin's rott-weiler" (recalling* THOMAS *"Darwin's Bulldog"* HUXLEY*). Ardent atheist. The Atheist Alliance organization instituted the Richard Dawkins Award in his honour in 2003. There can be no doubt that the similarity of the names Dawkins and Darwin unconsciously drives Dawkins's scientific ambitions.*

"I am against religion because it teaches us to be satisfied with not understanding the world."

"Modern theists might acknowledge that, when it comes to Baal and the Golden Calf, Thor and Wotan, Poseidon and Apollo, Mithras and Ammon Ra, they are actually atheists. We are all atheists about most of the gods that humanity has ever believed in. Some of us just go one god further."

On God as a "meme," Dawkins's idea of a unit of cultural inheritance, analogous to the gene: "How does it replicate itself? By the spoken and written word, aided by great music and great art. . . . The survival value of the god meme results from its great psychological appeal. It provides a superficially plausible answer to deep and troubling questions about existence. It suggests that injustices in this world may be rectified in the next. It holds out a cushion against our own inadequacies which, like a doctor's placebo, is none the less effective for being imaginary. There are some of the reasons why the idea of God is copied so readily by successive generations of individual brains."

"The meme for blind faith secures its own perpetuation by the simple unconscious expedient of discouraging rational inquiry."

"If complex organisms demand an explanation, so does a complex designer. And it's no solution to raise the theologian's plea that God (or the Intelligent Designer) is simply immune to the normal demands of scientific explanation. . . . You cannot have it both ways. Either ID belongs in the science classroom, in which case it must submit to the discipline required of a scientific hypothesis. Or it does not, in which case get it out of the science classroom and send it back into the church, where it belongs."

"The creationists' fondness for 'gaps' in the fossil record is a metaphor for their love of gaps in knowledge generally. Gaps, by default, are filled by God."

"Faith is powerful enough to immunize people against all appeals to pity, to forgiveness, to decent human feelings. It even immunizes them against fear, if they honestly believe that a martyr's death will send them straight to heaven. What a weapon! Religious faith deserves a chapter to itself in the annals of war technology, on an even footing with the longbow, the warhorse, the tank, and the hydrogen bomb."—*1976*

"Could we get some otherwise normal humans and somehow persuade them that they are not going to die as a consequence of flying a plane smack into a skyscraper? . . . The afterlife-obsessed suicidal brain really is a weapon of immense power and danger. It is comparable to a smart missile. . . . Yet . . . it is very very cheap. . . . To fill a world with religion, or religions of the Abrahamic kind, is like littering the streets with loaded guns. Do not be surprised if they are used."—*2001*

"[A letter to a U.K. newspaper] says 'science provides an explanation of the mechanism of the [December 2004 Asian] tsunami but it cannot say why this occurred any more than religion can.' There, in one sentence, we have the religious mind displayed before us in all its absurdity. In what sense of the word 'why', does plate tectonics not provide the answer? Not only does science know why the tsunami happened, it can give precious hours of warning. If a small fraction of the tax breaks handed out to churches, mosques and synagogues had been diverted into an early warning system, tens of thousands of people, now dead, would have been moved to safety. Let's get up off our knees, stop cringing before bogeymen and virtual fathers, face reality, and help science to do something constructive about human suffering."

Democritus (c. 460–370 B.C.E.), *Greek philosopher. Originator of the atomic view of matter. Even thoughts, morality, the soul were products of arrangements and interactions of atoms. All was thus material, nothing*

"spiritual." Was among the first to propose that the universe contains many worlds, some of them inhabited.

"Nothing exists except atoms and empty space; everything else is opinion."

Daniel Dennett *(1942–), American philosopher. Leading proponent of the theory that human intelligence and consciousness can be explained by Darwinian selection. His book* Breaking the Spell: Religion as a Natural Phenomenon *proposes evolutionary explanations for religious belief.*

"[ALFRED RUSSELL] WALLACE, the co-discoverer of natural selection, said that it covered everything up to the human soul and he drew the line there, exactly where DESCARTES drew the line. . . . To Darwin it was clear that the Cartesian line was indefensible simply because it was clear that we're primates—we're mammals. The continuity of nature was not going to permit one species on the planet to have miracle stuff in its brain when no other species did."

"I was once interviewed in Italy and the headline of the interview the next day was wonderful. . . . 'Yes, we have a soul, but it's made of lots of tiny robots,' and I thought, it's exactly right. Yes! We have a soul, but . . . it's mechanical . . . but it's still a soul. It still does the work that the soul was supposed to do. It is the seat of reason . . . of moral responsibility. It's why we are appropriate objects of punishment when we do evil things [and of] praise when we do good things. It's just not a mysterious lump of wonder-stuff. . . . Our moral quandaries [and] aspirations are what they were before. Our capacity to love or to hate are intact. . . . Darwinism changes everything and leaves everything the same."

René Descartes *(1596–1650), French philosopher; mathematician; scientist. The first philosopher to reject any appeal to* ends, *divine or natural, in explaining natural phenomena. The Catholic Church placed his works on the* Index of Prohibited Books *a few years after his death.*

Most people don't know the half of it: "Dubito ergo cogito, cogito ergo sum. [I doubt therefore I think, I think therefore I am.]"

John Dewey (1859–1952), *American pragmatist philosopher, educational reformer, social activist. Sought laws to protect minorities, legalize labor unions, and curb business monopolies; called the members of Congress "errand boys of big business." Regarded nature as the only reality, and values and beliefs as "products of human experience in nature." But repudiated what he called militant atheism.*

"Intellectually, religious emotions are not *creative* but *conservative*. They attach themselves readily to the current view of the world and consecrate it."

"Apologists for a religion often point to the shift that goes on in scientific ideas as evidence of the unreliability of science. . . . Even if the alleged unreliability were as great as they assume (or even greater), the question would remain: Have we any other recourse for knowledge? But in fact they miss the point. Science is not constituted by any particular body of subject matter. It is constituted by a method, a method of changing beliefs by means of tested inquiry. . . . The scientific-religious conflict ultimately is a conflict between allegiance to this method and allegiance to [any] belief so fixed in advance that it can never be modified."

Diagoras "the Atheist" of Melos (fifth century B.C.E.), *Greek poet. Became an atheist after a man who perjured against him went unpunished by the gods. When he was shown votive pictures of people reportedly saved from storms at sea by making vows to the gods, he replied, "there are nowhere any pictures of those who have been drowned." When he found himself aboard a ship in a dangerous storm and the crew thought they had brought it on themselves by taking this blasphemer on board, Diagoras asked if the other boats out in the same storm also had a Diagoras on board. Once threw a wooden image of a god into a fire, remarking that the deity should perform another miracle and save itself. The* hullabaloo *(Greek for "brouhaha") this caused in Athens forced him to flee for his life. Athens offered a reward for his capture dead or alive. Lived out his life in Spartan territory, presumably on a Spartan diet.*

Charles Dickens (1812–1870), *English novelist. Prayed regularly and accepted the teachings of Jesus, but rejected his divinit; hated the Roman Catholic Church, the Church of England, and the latter's influence in politics; and ridiculed evangelicals repeatedly in his novels.*

> "I cannot sit under a clergyman who addresses his congregation as though he had taken a return ticket to heaven and back."

Emily Dickinson (1830–1886), *American poet. Refused to make the public confession of faith required to formally join her church. Was "brilliant at keeping the tension of doubt, and at generating a private religion, of art and inner life, that 'doubts as fervently as it believes.'"* [2]

> "'Faith' is a fine invention, when gentlemen can see / But microscopes are prudent, in an emergency."

> "Those—dying then, / Knew where they went—/ They went to God's Right Hand— / That Hand is amputated now / And God cannot be found. . . ."

Denis Diderot (1713–1784), *French philosopher, author, encyclopedist. His parents, wishing him to be a priest, shaved his head and dubbed him "abbé" (abbott) at age 13. During his university studies in science and philosophy, he began his rise from faith to deism to the highest stage, atheism. His book* Pensées Philosophiques, *which expressed anti-Christian views, was condemned and burned by the public executioner. Jailed for questioning the existence of God in his next book,* An Essay on Blindness. *Theorized —100 years B.D. (before Darwin)—that animal species had evolved from common ancestors.*

> "I have only a small flickering light to guide me in the darkness of a thick forest. Up comes a theologian and blows it out."

> "The philosopher has never killed any priests, whereas the priest has killed a great many philosophers."

> "The man who first pronounced the barbarous word God ought to have been immediately destroyed."

"The Christian religion: the most absurd in its dogmas, the most unintelligible, the most insipid, the most gloomy, the most Gothic, the most puerile."

"It is very important not to mistake hemlock for parsley, but to believe or not believe in God is not important at all."

From his "Conversation with a Christian Lady":

"Then you're the man who doesn't believe anything?. . . . Yet your moral principles are the same as those of a believer. . . . What? You don't steal? You don't kill people? You don't rob them?. . . . Then what do you gain by not being a believer?" "Nothing at all, madame. Is one a believer from motives of profit?"

Marlene Dietrich (1901–1992), *German-born actress and singer. Became a U.S. citizen in 1937. Staunch anti-Nazi. During the 1980s she reportedly ran up a $3,000 monthly phone bill talking to world leaders, including Ronald Reagan and Mikhail Gorbachev.*

"If there is a supreme being, he's crazy." *(That must be how you feel after talking to supreme leaders like Reagan.)*

Ellen Battelle Dietrick (d. 1895), *American suffragist and campaigner for church-state sep.*

"The first chapter of Genesis . . . tells us [in Hebrew], in verses one and two . . . '[In the beginning] created *the gods (Elohim)* these skies . . . and this earth. . . .' Here we have the opening of a polytheistic fable of creation, but, so strongly convinced were the English translators that the ancient Hebrews must have been originally monotheistic that they rendered the above as follows: 'In the beginning *God* created the heaven and the earth. . . .'" *(Italics added.)*

"There is a tree of evil, whose fruit is said by Iahveh [Big G] to cause sudden death, but which does not do so, as Adam lived 930 years after eating it."

Annie Dillard (1945–), *American novelist/poet/nonfiction writer.*

"I read about an Eskimo hunter who asked the local missionary priest, 'If I did not know about God and sin, would I go to hell?' 'No,' said the priest, 'not if you did not know.' 'Then why,' asked the Eskimo, 'did you tell me?'"

Phyllis Diller (1917–), *American comedian. Having based her comedy largely on her self-proclaimed ugliness, she received numerous awards and tributes from plastic surgeons for achievement in facial remodeling. Holds a Guinness World Record for delivering 12 punchlines per minute.*

Asked how she visualized the hereafter: "There isn't any, you dingbat! . . . Ahhha . . . ha. . . . ha . . . ha . . . haaah. . . . This is it, baby!" *How ironic: After a heart attack in 1999, Diller was pronounced clinically dead for three minutes before recovering.*

"Religion is such a medieval idea. Don't get me started. . . . Aahh, it's all about money. . . ."

Diodorus Siculus (c. 90–30 B.C.E.), *Greek historian. His earliest recorded remarks concern an incident he witnessed during a visit to Egypt: an angry mob demanding the death of a Roman citizen who had accidentally killed a cat, an animal sacred to the ancient Egyptians.*

"It is in the interest of states to be deceived in religion. . . . The myths about Hades and the gods, although they are pure invention, help to make men virtuous." *As in the aforementioned mob in Egypt?*

Diogenes "the Cynic" (412–323 b.c.e.), *Greek philosopher. Yeah, the one who walked around with a flashlight looking for an honest man. Identified morality with austerity and pain. Lived in a tub. Destroyed his single wooden bowl upon seeing a poor boy drink from his hands. Alexander the Great, upon meeting the famous philosopher, asked if there was any favor he might do for him. "Stand out of my sunlight," Diogenes*

said. When Plato (citing Socrates) defined man as a "featherless biped," Diogenes plucked a chicken, brought it into Plato's school, and said, "This is Plato's man." Okay, so you knew all that.

Upon crushing a louse on the altar rail of a temple: "Thus does Diogenes sacrifice to all the gods at once." *They don't call them Cynics for nothing.*

"When I look upon seamen, men of science, and philosophers, man is the wisest of all beings. When I look upon priests and prophets . . . nothing is so contemptible as man."

Benjamin Disraeli *(1804–1881), Jewish-born English politician and novelist. Served twice as prime minister and nearly four decades in the House of Commons. At age 13, instead of a bar mitzvah, he was baptized an Anglican following a dispute his father had with their synagogue. It's considered one of the nineteenth century's most imaginative excuses for conversion out of Judaism. (Compare and contrast, e.g.,* HEINRICH HEINE.)

"Where knowledge ends, religion begins."

"Man is made to adore and obey: but if you will not command him, if you give him nothing to worship, he will fashion his own divinities."

Theodosius Dobzhansky *(1900–1975), Ukrainian-born geneticist and evolutionary biologist. Key figure in the modern synthesis of evolutionary biology with genetics. Author of a famous anti-creationist essay, "Nothing in Biology Makes Sense Except in the Light of Evolution."*

"One of the early antievolutionists, P. H. Gosse, published a book [in 1857] entitled *Omphalos* ('the Navel'). The gist of this amazing book is that Adam, though he had no mother, was created with a navel, and that fossils were placed by the Creator where we find them now—a deliberate act on His part, to give the appearance of great antiquity and geologic upheavals. . . . Such notions are blasphemies, accusing God of absurd deceitfulness."

"What is the sense of having as many as 2 or 3 million species living on earth? . . . Was the Creator in a jocular mood when he made *Psilopa petrolei* for California oil fields and species of *Drosophila* to live exclusively on some body-parts of certain land crabs on only certain islands in the Caribbean? . . . All this is understandable in the light of evolution theory; but what a senseless operation it would have been, on God's part, to fabricate a multitude of species ex nihilo and then let most of them die out!"

E. L. (Edgar Lawrence) Doctorow (1931–), *American novelist.*

"Let us celebrate the constancy of the speed of light, let us praise gravity, that it is in action the curvature of space, and glory that even light is bent by its force, riding the curvatures of space toward celestial objects as a fine, shimmering red-golden net might drape over them."

Phil Donahue (1935–) *American talk-show host.* The Phil Donahue Show *was the first, and longest-running, tabloid talk show—27 years (1969–1996). When his brief comeback show (2002–2003) was canceled, it was MSNBC's highest-rated. A leaked NBC memo said Donahue—an Iraq-war opponent and one of the few liberal voices on national TV—would be a "difficult public face for NBC in a time of war." "[A] pious atheist, a viewpoint represented nowhere in the cable news media," the* American Prospect *noted. Attended an all-boys Catholic prep school run by the Brothers of Holy Cross. That'll usually do it. From his 1985 book* The Human Animal:

"Science may have come a long way, but as far as religion is concerned, we are first cousins to the !Kung tribesmen of the Kalahari Desert. Except for the garments, their deep religious trances might just as well be happening at a revival meeting or in the congregation of a fundamentalist TV preacher. . . . As we move further from the life of ignorance and superstition in which religion has its roots, we seem to need it more and more. . . . Why has religion become a force just when we'd have thought it would be losing ground to secularism?"

Hundred-thousand-year-old habits die hard? The human race will always remain a faithoholic, on the wagon or off? "Superstition is rooted in a much deeper and more sensitive layer of the psyche than skepticism" (GOETHE)? PASCAL's Wager? A hundred years of commercialism and advertising has eroded our sense of reality and truth, while the products themselves fail to fill the spiritual void? The world has become so ugly, so mechanized, paved, the culture so crappy, and life so literally disenchanted, that people turn to hallowed myths and to shared emotions and music (see next entry), less for truth or meaning than for beauty? . . . Just trying to help.

Amanda Donohoe (1962–), *English actress. Devout feminist and socialist. Once punched out an obsessed fan of her pop-singer boyfriend Adam Ant when the fan held him at knifepoint on British TV. (Why are all the good ones taken?) On her role as a pagan priestess in the film* Lair of the White Worm, *in which she spat venom onto a crucifix:*

> "I'm an atheist, so it was actually a joy. Spitting on Christ was a great deal of fun.* I can't embrace a male god who has persecuted female sexuality throughout the ages."

Fyodor Dostoyevsky (1821–1881), *Russian author. Following years of hard labor in Siberia for engaging in liberal political activity, became a Christian and conservative. "The quest for God, the problem of Evil and suffering of the innocents haunt the majority of his novels."* [1] *WALTER KAUFMANN called Dosty's novel* Notes from Underground *"the best overture for existentialism ever written." NIETZSCHE said of JESUS, "it is regrettable that no Dostoyevsky lived near him." Compulsive gambler, wrote* Crime and Punishment *quickly to pay off a gambling debt. What version of poker would Jesus play?*

> "So long as man remains free he strives for nothing so incessantly and so painfully as to find something to worship. But . . . what is essential is that all may be *together* in it. This craving for *community* of worship is the chief misery of every man individually and of all humanity from the beginning of time. For

---

* *The Quotable Atheist* urges respect for religious symbols. Burn them, bury them, but do not spit on them.

the sake of common worship, they've slain each other with the sword. They have set up gods and challenged each other, 'Put away your gods and come and worship ours, or we will kill you and your gods!'"

Frederick Douglass (1818–1895), *African-American abolitionist leader. Critics of his* Narrative of the Life of Frederick Douglass, an American Slave—*a bestseller reprinted nine times in its first three years—refused to believe a black man could have written so eloquent a book. His marriage to Helen Pitts, a white woman 20 years his junior, scandalized the country and both partners' families. Her abolitionist parents—former friends of Douglass—stopped speaking to her.*

"I prayed for freedom for twenty years, but received no answer until I prayed with my legs."

"The church of this country is not only indifferent to the wrongs of the slave, it actually takes sides with the oppressors. . . . For my part, I would say, welcome infidelity! Welcome atheism! Welcome anything! in preference to the gospel, as preached by these Divines! They convert the very name of religion into an engine of tyranny and barbarous cruelty, and serve to confirm more infidels, in this age, than all the infidel writings of THOMAS PAINE, VOLTAIRE, and Bolingbroke put together have done!"

"We have men sold to build churches, women sold to support the gospel, and babes sold to purchase Bibles for the poor heathen, all for the glory of God and the good of souls. The slave auctioneer's bell and the church-going bell chime in with each other, and the bitter cries of the heart-broken slave are drowned in the religious shouts of his pious master. Revivals of religion and revivals in the slave trade go hand in hand."

John William Draper (1811–1882), *English-born American chemist, historian, and photographer. Professor at New York University. First person to take an astrophotograph (the moon, 1840) and a deep space photograph (Orion nebula, 1880). Coauthor of the "conflict theory," which holds that religion and science are almost always at odds.*

"The history of science is not a mere record of isolated discoveries; it is a narrative of the conflict of two contending powers, the expansive force of the human intellect on the one hand, and the compression arising from traditionary faith and human interest on the other."

"Though there is a Supreme Power, there is no Supreme Being. There is an invisible principle, but not a personal God. . . . All revelation is, necessarily, a mere fiction. That which men call chance is only the effect of an unknown cause. Even of chances there is a law . . . the universe is only a vast automatic engine. The vital force which pervades the world is what the illiterate call God."

Theodore Dreiser (1871–1945), *American writer. Best known for the novels* Sister Carrie *and* An American Tragedy.

"[Religion] is a bandage that man has invented to protect a soul made bloody by circumstance."

"Assure a man that he has a soul and then frighten him with old wives' tales as to what is to become of him afterward, and you have hooked a fish, a mental slave."

Ann Druyan (1949–), *American author and media producer. Wife of the late* CARL SAGAN. *President of the NORML (National Organization for the Reform of Marijuana Laws) Foundation Board of Directors. (God might not be in every plant, but certainly in some.) Selected the music portion of the "Golden Record" of information about Earth sent into space in 1977 on board the two Voyager space probes. (Sagan headed the committee that designed the record for NASA and that was forced to replace a photograph of a naked man and woman with silhouettes because of criticism of the "smut" by you know who, basically.) Has been called a pantheist (and worse).*

"The roots of this antagonism to science run very deep. . . . We see them in Genesis . . . in which the first humans are doomed and cursed eternally for asking a question, for partaking of the fruit of the Tree of Knowledge. . . . [Eden] is more like a maximum-security prison with twenty-four hour surveillance. It's a horrible place."

W. E. B. DuBois (1868–1963), *African-American historian, civil rights activist, communist, rationalist, agnostic. Helped establish the NAACP. Charged with being an unregistered agent of the Soviet Union during the McCarthy witch-hunt.*

"I think the greatest gift of the Soviet Union to modern civilization was the dethronement of the clergy and the refusal to let religion be taught in the public schools."

"The theology of the average colored church is basing itself far too much upon 'hell and damnation.' . . . Our present method of periodic revival [involves] the hiring of professional and loud-mouthed evangelists and reducing people to a state of frenzy or unconsciousness."

Finley Peter Dunne (1867–1936), *American writer/humorist, best known for his nationally syndicated "Mr. Dooley" newspaper pieces, in which a Chicago Irish bar owner brogued away on issues of the day. "Dooley" was read aloud at Theodore Roosevelt's cabinet meetings. Dunne's wife, Margaret Abbott, was the first American woman to win an Olympic gold medal (golf).*

"A fanatic is a man who does what he thinks the Lord would do if He knew the facts of the case."

John J. Dunphy, *American haiku protest poet and writer. From "A Religion for a New Age," his award-winning 1983 essay for* The Humanist *magazine, which inspired hate-mail from the religious right and has often been quoted as evidence of the Vast Secular-Humanist Conspiracy Against God (to which I hope you remembered to pay this year's dues):*

"The battle for humankind's future must be waged and won in the public school classroom by teachers who correctly perceive their role as the proselytizers of a new faith: a religion of humanity that recognizes and respects the spark of what theologians call divinity in every human being. . . . The classroom must and will become an arena of conflict between the old and the new—the rotting corpse of Christianity, together with all its [attendant] evils and misery, and the new faith of humanism, resplendent with the promise of a world in which the Christian ideal of 'love thy neighbor' will finally be achieved."

Will Durant (1885–1981), *American philosopher and atheist. Author of the huge and hugely popular* Story of Philosophy (1926). *His study of science and philosophy at a Jesuit seminary destroyed his religious beliefs. Excommunicated soon afterward for giving a lecture about the subcurrent of sex in religion and the worship of the phallus in earlier cultures as a symbol of divine power. At age 28 he married a 15-year-old student of his who eventually became his coauthor.*

> *1932:* "The greatest question of our time is not communism vs. individualism, not Europe vs. America, not even the East vs. the West; it is whether men can bear to live without God."

> "Intolerance is the natural concomitant of strong faith; tolerance grows only when faith loses certainty; certainty is murderous."

Emile Durkheim (1858–1917), *French philosopher and one of the founders of modern sociology. Father and grandfather were rabbis. Secularist and socialist himself. Much of his work sought to demonstrate that religious and moral beliefs were determined by social, not divine, factors.*

> "Sacred things are simply collective ideals that have fixed themselves on material objects."

Freeman Dyson (1923–) *British-born American physicist, mathematician, and Princeton professor. Led the Orion Project (1957–1961) to design a nuclear-propelled U.S. spacecraft. Calls himself "a skeptical Christian." Awarded the Templeton Prize for oxymoronic Progress in Religion.*

> *Don't say we don't offer choices:* "There are two kinds of atheists—ordinary atheists who do not believe in God and passionate atheists who consider God to be their personal enemy." *[His prime example of the latter:* PAUL ERDÖS.*]*

Umberto Eco (1932–), *Italian writer. Best known for his novel* The Name of the Rose, *in which a scientific-minded medieval monk/detective solves a murder case. Left the church at age 22. Had literally not moved from his pew in over 14 years. (Not true.)*
An agnostic who declares:

> "One should not have the arrogance to declare that God does not exist."

But also:

> "Fear prophets and those prepared to die for the truth, for as a rule they make many others die with them, often before them, at times instead of them."

Sir Arthur Eddington (1882–1944), *British astrophysicist and science popularizer. His observations were hailed as conclusive proof of EIN-STEIN's theory of general relativity. (When a reporter asked if it was true that only three people in the world understood relativity, he replied, "Oh, who's the third?") Developed the first true understanding of what goes on inside stars. (You don't want to know.) His estimate of the number of electrons in the universe (1.56×10^{79}) is known as the Eddington Number. As a Quaker and pacifist, refused to serve in the military in WWI.*

"We are bits of stellar matter that got cold by accident, bits of a star gone wrong."

Taner Edis (1967–), *Turkish-born American physicist and author. Professor of physics at Truman State University in Missouri who clearly lives in a "show me" state of mind. Coeditor of* Science and Nonbelief *(2006) and* Why Intelligent Design Fails *(2004); author of* The Ghost in the Universe: God in Light of Modern Science *and of numerous articles (including several about Islamic creationism) published in the "freethinking" press.*

"Physicists use 'God' as a metaphor more often than other scientists. . . . Of course, this is just a metaphor for order at the heart of confusion. A rational or aesthetic pattern underlying reality is far from a theistic God."

"Theologians hardly predicted the Big Bang. . . . Even if a beginning for the universe is a successful prediction of one version of theism, this is still not that impressive. After all, even a stopped clock is right twice a day. The Big Bang becomes strong support for God only with an argument showing that such a beginning requires a Creator."

"Creation out of absolute nothing is a metaphysical quagmire for theists anyway, since nothing must at least have the potentiality for becoming something. Since theists are stuck with potentiality, it might as well be something like a quantum vacuum [*a state which by definition contains no physical particles*]. . . . Quantum events have a way of just happening, without any cause. . . . Even the quantum vacuum is not an inert void, but is boiling with quantum fluctuations. Energy fluctuations out of nothing create short-lived particle-antiparticle pairs. . . . An uncaused beginning, even out of nothing, for spacetime is no great leap of the imagination. . . . In all likelihood, the universe is uncaused. . . . *It just is.*"

Thomas Edison (1847–1931), *American inventor (electric light, phonograph, movie projector, telephone transmitter, dozens of other amusing*

but impractical contraptions). Attended school for only three months. Agnostic or atheist from boyhood. When a minister asked about installing lightning rods atop his church, Edison answered: "By all means, as Providence is apt to be absent-minded." Hoping to influence him, his wife, a devout Methodist, would invite clergymen to dinner—six bishops at once on one occasion. Asked about immortality, he pointed to an electric light and said: "There lives Thomas Edison." In 1920, the mischievous imp of Menlo Park announced he was working on an electronic device for communicating with departed souls. Was denounced from the pulpits for remarks like these:

"Religion is all bunk."

Knew an invention when he saw one: "All Bibles are man-made."

"The great trouble is that the preachers get the children from six to seven years of age, and then it is almost impossible to do anything with them. Incurably religious—that is the best way to describe the mental condition of so many people. Incurably religious."

Paul Edwards (born Paul Eisenstein, 1923–2004), *Austrian-Jewish-American philosopher. Edited and wrote the introduction to longtime friend* BERTRAND RUSSELL'S Why I Am Not a Christian. *Follower of Wilhelm Reich's "primal scream" therapy. Possessed one of Reich's "orgone accumulators." (A lot simpler than relationships. Two D batteries and you're ready to partay.)*

"Atheism may be defined as the view that 'God exists' is a false statement. But there is also a broader sense in which an atheist is someone who rejects belief in God. . . . It may be rejected because it is incoherent or meaningless, because it is too vague to be of any explanatory value, or because, as LAPLACE put it in his famous exchange with Napoleon, there is no need for this 'hypothesis'. Atheism in this broader sense remains distinct from agnosticism, which advocates suspension of judgement. It is surely possible to justify atheism in this broader sense."

Greg Egan (1961–), *Australian science-fiction author. Described his Hugo Award-winning novella* Oceanic *as a thinly veiled account of his*

own journey to atheism, and his story "The Moral Virologist" as "a direct response to religious fundamentalists blathering on about AIDS being God's instrument. . . . This was a blasphemous obscenity."

"Most Christian theologians have retreated from all the things that their religion supposedly asserts; they take a much more 'modern' view than the average believer. But by the time you've 'modernised' something like Christianity—starting off with 'Genesis was all just poetry' and ending up with 'Well, of course there's no such thing as a personal God'—there's not much point pretending that there's anything religious left. You might as well come clean and admit that you're an atheist with certain values, which are historical, cultural, biological, and personal in origin, and have nothing to do with anything called God."

Barbara Ehrenreich (1941–), *American journalist/essayist/ progressive social critic. Holds a BA in physics and a Ph.D. in cell biology. Her 2001 book* Nickel and Dimed, *about her attempt to live on low-wage jobs, sold over a million copies. (Lots of nickels and dimes.) Her most gripping book, however, was* The Uptake, Storage, and Intracellular Hydrolysis of Carbohydrates by Macrophages *(1969).*

About her waitressing job: "The worst [customers], for some reason, are the Visible Christians—like the ten-person table, all jolly and sanctified after Sunday night service, who run me mercilessly and then leave me $1 on a $92 bill. Or the guy with the crucifixion T-shirt . . . who complains that his baked potato is too hard and his iced tea too icy (I cheerfully fix both) and leaves no tip at all. As a general rule, people wearing crosses or WWJD ('What Would Jesus Do?') buttons look at us disapprovingly no matter what we do, as if they were confusing waitresses with Mary Magdalene's original profession."

"God has a lot to account for in the way of earthquakes, hurricanes, tornadoes, and plagues. Nor has He ever shown much discrimination in his choice of victims. A tsunami hit Lisbon in 1755, on All Saints Day, when the good Christians were all in church. The faithful perished, while the denizens of the red light district, which was built on strong stone, simply carried on sinning. Similarly, last fall's hurricanes flattened the God-fearing,

Republican parts of Florida while sparing sin-soaked Key West and South Beach." *(Also see CHARLES FIELD.)*

On the December 2004 Asian tsunami:

"The Christian-style 'God of love' should be particularly vulnerable to post-tsunami doubts. . . . If He so loves us . . . why couldn't he have held those tectonic plates in place at least until the kids were off the beach? . . . If we are responsible for our actions, as most religions insist, then God should be, too, and I would propose an immediate withdrawal of prayer and other forms of flattery . . . at least until an apology is issued."

Albert Einstein (1879–1955), *German-Jewish-American physicist. His famous remark "God does not play dice" referred to quantum mechanics and the role of chance in physics—but was widely misinterpreted—as this first quote—a response to a letter from a worried atheist, indicates:*

"It was, of course, a lie what you read about my religious convictions, a lie which is being systematically repeated. I do not believe in a personal God and I have never denied this but have expressed it clearly. If something is in me which can be called religious then it is the unbounded admiration for the structure of the world so far as our science can reveal it."

"[The sense of] a spirit manifest in the laws of the Universe. . . . does not lead us to take the step of fashioning a god-like being in our own image—a personage who makes demands of us and who takes an interest in us as individuals. There is in this neither a will nor a goal, nor a must, but only sheer being." *(Bishop Fulton J. Sheen said about Einstein's "faith": "Who ever wanted to die for the Milky Way?" Exactly. Or kill for it, he might have added.)*

"[Religion is] an attempt to find an out where there is no door."

"A man's ethical behavior should be based effectually on sympathy, education, and social ties and needs; no religious basis is necessary. If people are good only because they fear punishment and hope for a reward, then we are a sorry lot indeed."

"Only two things are infinite: the universe and human stupidity, and I'm not sure of the former."

Does religion promote peace? "It has not done so up to now."

Riane Eisler (1931–), *Austrian-Jewish-Cuban-American attorney, author, social historian, and "cultural transformation" activist. Everything you could want in a woman. Her family escaped Nazi Austria after Kristalnacht, first to Cuba, then to the U.S. Advocates a return from our allegedly male-dominated "dominator culture" or "androcracy" to the alleged "partnership model" of ancient cultures. Author of* The Chalice and the Blade *and* Sacred Pleasure: Sex, Myth, and the Politics of the Body.

> *Refering to prehistoric "Venus" figurines:* "If the central religious image [in Neolithic times] was a woman giving birth and not, as in our time, a man dying on a cross, it would not be unreasonable to infer that life and the love of life—rather than death and the fear of death—were dominant in society as well as art." *(But see* ANGELA CARTER.*)*

Steve Eley, *self-anointed "Chief Advocate and Spokesguy" of the religion of the Invisible Pink Unicorn (IPU, "Blessed Be Her Holy Hooves"), whose earliest written documents are from the Usenet discussion group alt.atheism in the early 1990s C.E. Despite rivals such as the Cult of the Very Stealthy Maroon Pegasus,* CARL SAGAN's *Invisible Green Dragon, and* BOBBY HENDERSON's *Church of the Flying Spaghetti Monster, IPU has become an emblem for atheists, with its logo emblazoned on T-shirts, coffee mugs, etc. (One might suppose the IPU, were She to return, would drive the merchants from the Web sites.)*

> "Invisible Pink Unicorns are beings of great spiritual power. We know this because they are capable of being invisible and pink at the same time. Like all religions, the Faith of the Invisible Pink Unicorns is based upon both logic and faith. We have faith that they are pink; we logically know that they are invisible because we can't see them."

George Eliot (born Mary Anne Evans, 1819–1880), *English novelist. Born into a pious family, her childhood faith was shattered by reading* SIR WALTER SCOTT. *Refused to accompany her family to church. Was soon consorting with freethinkers like* RALPH WALDO EMERSON *and* ROBERT

OWEN. *Because of her agnosticism, Eliot was denied a burial spot in the exclusive Poet's Corner of Westminster Abbey after her death.*

> "Your dunce who can't do his sums always has a taste for the infinite."

> "God, immortality, duty—how inconceivable the first, how unbelievable the second, how peremptory and absolute the third."

> "My childhood was full of deep sorrows—colic, whooping-cough, dread of ghosts, hell, Satan, and a Deity in the sky who was angry when I ate too much plumcake." *The Bible is quite clear on that question.*

On evangelism as a career alternative:

> "Given a man with moderate intellect, a moral standard not higher than the average, some rhetorical affluence and a great glibness of speech, what is the career in which, without the aid of birth or money, he may most easily attain power and reputation in English society? Where is that Goshen of mediocrity in which a smattering of science and learning will pass for profound instruction, where platitudes will be accepted as wisdom, bigoted narrowness as holy zeal, unctuous egoism as God-given piety?"

Elisha ben Abuyah (ca. 70–135 c.e.), *Palestinian rabbi turned atheist. Nicknamed Aher, "the Other," and eventually excommunicated. Lover of Greek poetry and philosophy. The Talmud tells how, while a rabbinical student, he hid forbidden books in his clothes, and how a voice from heaven once called out: "Turn, O backsliding children, with the exception of Aher." (Alternately translated as: "Turn, O backsliding children. Not you, Aher.")*

The declaration that led to his excommunication, uttered upon seeing a child fall from a tree and die while attempting to perform a mitzvah, a religious duty: "There is no justice, and there is no Judge."

Havelock Ellis (1859–1939), *British physician, sexual psychologist, and social reformer. Author of the seven-volume* Studies in the Psychology of Sex, *which until 1935 was legally available only to the medical profession. Coauthor of the first English medical textbook on homosexuality, which he did not consider a disease, sin, or crime.*

"The whole religious complexion of the modern world is due to the absence from [ancient] Jerusalem of a lunatic asylum."

Harlan Ellison (1934–), *award-winning American author of science fiction, horror and mystery stories, essays, and criticism. Wrote for the original* Outer Limits *and* Star Trek *TV series. Dropped out of Ohio State University, where a professor had told him he would never be a writer. "For the next forty-and-some years, he sent this man a copy of every article, story, and book he turned out."[1] "His story 'The Deathbird' presents an alternate take on the biblical account of creation, wherein the snake is a Prometheus figure sent to give humanity the wisdom it will need to overcome the tyrannical rule of the insane being who calls himself God."[4]*

> *Not a moron:* "No, I don't believe in God. . . . I'm not a moron. I have to have some proof of something."

> "God [*who doesn't exist*] has more important things to do than talk to little French girls in jail [*i.e., Joan of Arc*] . . . and give you hair growing on the palm on your hand if you masturbate."

Ralph Waldo Emerson (1803–1882), *American author, poet, philosopher. A Unitarian minister, like his father and grandfather—until, as his wife said, he "left his pulpit as a matter of honor." First branded an infidel and atheist for proclaiming in a Harvard Divinity School graduation address that Jesus was not God. Formulated the philosophy of Transcendentalism (also see* THOREAU*), which rejected religion yet saw divinity throughout nature and postulated an intuitional link between individual human psyches and a great, mystical Over-Soul. Or as some smartass poet put it: "Many converts they've got—to I don't, nor they either, exactly know what."*

> "As men's prayers are a disease of the will, so are their creeds a disease of the intellect."

> "An actually existent fly is more important than a possibly existent angel."

> "We are born believing. A man bears beliefs, as a tree bears apples."

> "The religion of one age is the literary entertainment of the next."

> *Cynical sociologism:* "The god of the cannibals will be a cannibal,

of the crusaders a crusader, and of the merchants a merchant."

"The religion that is afraid of science dishonors God and commits suicide."

Warning to Intelligent Design-ists: "Don't set out to teach theism from your natural history. . . . You'll spoil both."

"We must get rid of that Christ."

"Every man is a divinity in disguise, a god playing the fool."

"The dull pray; the geniuses are light mockers."

"Nothing is at last sacred but the integrity of your own mind."

"I hate quotations. Tell me what you know."

Empedocles (ca. 490–430 B.C.E.), *Greek philosopher. Originator of the four-element theory of matter. Postulated a crazy little thing called Love (philia) to explain the attraction of different forms of matter. One of the first to theorize that light travels at a finite speed. Claimed that by the virtue of his knowledge, he had become divine (we freethinkers get like that), but, in keeping with his democratic politics, believed others could become divine too.*

"None of the gods has formed the world, nor has any man; it has always been."

Friedrich Engels (1820–1895), *German-British industrialist and socialist. MARX's angel, as it were; coauthor of* The Communist Manifesto.

"What, indeed, is agnosticism, but, to use an expressive term, 'shamefaced' materialism."

Epicurus (341–270 B.C.E.), *Greek philosopher. Taught that the soul consists of atomic material that disintegrates at death.*

Makes short work of Him: "Is God willing to prevent evil, but not able? Then he is not omnipotent. Is he able, but not willing? Then he is malevolent. Is he both able and willing? Then whence cometh evil? Is he neither able nor willing? Then why call him God?"

"If God listened to the prayers of men, all men would quickly have perished: for they are forever praying for evil against one another."

Paul Erdös (1913–1996), *Hungarian-born American mathematician —one of the most brilliant and eccentric of the past century. Could multiply three-digit numbers in his head at age three. Homeless for more than 20 years, appearing with a tattered suitcase at the home of a fellow mathematician, working together for a few days, then moving on to another. Right to the end, worked 19-hour days, fueled by pills and espresso. ("A mathematician," he said, "is a machine for turning coffee into theorems.")*

"I kind of doubt He [exists]. Nevertheless, I'm always saying that the SF [*Supreme Fascist—Erdös's customary name for G-d*] has this transfinite Book . . . that contains the best proofs of all theorems, proofs that are elegant and perfect. . . . You don't have to believe in God, but you should believe in the Book." *(i.e., we don't create mathematical truths—they're out there, waiting to be discovered.)*

"The SF created us to enjoy our suffering. The sooner we die, the sooner we defy His plans."

Susan Ertz (1894–1985), *British novelist.*

"Millions long for immortality who don't know what to do on a rainy afternoon."

"Parsons always seem to be specially horrified about things like sunbathing and naked bodies. They don't mind poverty and misery and cruelty to animals nearly as much."

Greg Erwin (d. 1998), *former vice president, Humanist Association of Canada (motto: Cogito, ergo atheos sum). Maintained the International Atheistic Secular Humanist Conspiracy (Canada Division) Web site, which once featured a guide to getting excommunicated from the Catholic Church.*

"Nearly every human group has created something in the way of a religion, no two of which are the same. When something is based on reality, like mathematics or scientific medicine, groups of people independently arrive at the same answers. . . . This is one good way to tell the difference between shit and shinola."

"The kind of things that religious people offer as evidence for their brand of religion, they do not accept as evidence when proferred by adherents of other religions. Religions do not accept each others' miracles, revelations, prophets, or holy books. . . . In the absence of any convincing reason to accept one set of claims while rejecting the rest, the simplest conclusion is that they are all. . . ." *(Perhaps Erwin was struck by a thunderbolt before he could finish the . . .)*

Carl G. Estabrook, *visiting professor of intellectual history, University of Illinois; columnist for* CounterPunch.

"For over a thousand years, the tradition of the seven deadly sins, from late antiquity, formed the basis of Christian moral exhortation—not the Ten Commandments. . . . There were still in the sixteenth century quite well-informed Catholics who had never heard of the Commandments. . . . What prompted the revolution in moral theory was the rise of capitalism. . . . An entire structure of obedience is spun out of 'Honor thy father'. . . . The Commandments were wrenched from their historical context and twisted in an authoritarian direction. . . . The Ten Commandments in their proper historical context are a revolutionary manifesto, dedicated to the overthrow of traditional authority and religion . . . a 'Declaration of Independence of Liberated Israel' [*from Egypt*]. . . . They commend atheism in regard to the religion of the gods and anarchism in respect to the laws of the kings. . . . YHWH is not a god in the sense of the surrounding society. . . . Even an image of YHWH is forbidden—the only image of YHWH is humanity (Genesis 1:26). . . . The Ten Commandments rejected a society that claimed the power of life and death ('You shall not kill'). . . . The commandment against stealing is not about property. . . . These commandments are condemnations of the powerful who invaded households to steal concubines and slaves."

Euripides

Euripides (c. 480–406 B.C.E.), *Greek tragedian. Shattered conventions by portraying strong women characters and smart slaves and by satirizing mythological heroes.*

"Do we, holding that the gods exist, deceive ourselves with insubstantial dreams and lies, while random careless chance and change alone control the world?"

"I sacrifice to no god save myself—And to my belly, greatest of deities."

NONBELIEVER * HUMANIST * RATIONALIST * FREETHINKER * AGNOSTIC * GODLESS * HERETIC * ATHEIST * SECULAR * INFIDEL *

Fairness and Accuracy In Reporting (FAIR),

American media watchdog group. 1996:

> "'She had the dubious distinction of being known as America's most outspoken atheist,' NBC's Tom Brokaw said in introducing a jokey segment on MADALYN MURRAY O'HAIR, who has been missing for the past year. It's impossible to imagine Brokaw making light of the disappearance of someone who has the 'dubious distinction' of being a leader of America's Catholics or Jews—but atheists are assumed to be fair game.... That must be why NBC quoted a 'conservative Christian commentator' as saying of O'Hair: 'If she is indeed dead, then she's burning in the fires of hell.' Plenty of fundamentalist Christians believe that all Catholics burn in hell, but we doubt we'll see NBC quoting any of them the next time a pope dies." *Brokaw went on to say "There are no atheists in foxholes"* twice on NBC Evening News, *in 2001 and 2003.*

FaithLens, *the Evangelical Lutheran Church of America Web site.*
May 2006:

> "'Religion is becoming a new brand,' says a representative of Youth Intelligence, a trend-forecasting company. . . . Numerous

---

* The phrase was first used by WWII reporter Ernie Pyle.

Internet retail companies offer t-shirts, bags, belt buckles, and other clothing items sporting messages like 'Saved,' 'Buddha Rocks,' or 'Moses is my Homeboy.' . . . Celebrities and normal people alike are wearing their faith on their sleeves—literally. The reasons for this growing trend are many. Young people are searching for group identity and wearing Faith Fashion items makes them feel unique and part of something bigger at the same time. 'It's like wearing a band uniform,' says Melissa, a Baylor University student. . . . Obviously, Faith Fashion sells well." *Yeah, well, wait till our new line of atheist sportswear hits the shelves.*

Jerry Falwell (1933–), *American evangelical pastor, televangelist, and leading excrescence. His* National Liberty Journal *was the first news source to warn that a* Teletubbies *character,* Tinky Winky, *might be a secret agent of the homosexual conspiracy to destroy America. If there is such a thing as an atheist sacrament, it is the mockery of Jerry.*

"Christians, like slaves and soldiers, ask no questions."

"If you're not a born-again Christian, you're a failure as a human being."

"The idea that religion and politics don't mix was invented by the devil to keep Christians from running their own country."

James Feibleman (b. 1904), *American millionaire autodidact without formal training or degree who became head of Tulane University's philosophy department and wrote more than 36 influential books.*

"A myth is a religion in which no one any longer believes."

Jules Feiffer (1929–), *American cartoonist. His work ran in the* Village Voice *for 42 years. Won a Pulitzer for editorial cartooning in 1986.*

"Christ died for our sins. Dare we make his martyrdom meaningless by not committing them?"

Federico Fellini (1920–1993), *Italian movie director. Four best foreign film Oscars. Federico Fellini International Airport in Rimini, his hometown. Raised as a devout Catholic. Frequently consulted astrologers, mediums, and clairvoyants.*

> "Like many people, I have no religion, and I am just sitting in a small boat drifting with the tide. I live in the doubts of my duty. . . . I think there is dignity in this, just to go on working."

Francisco Ferrer (1859–1909), *Spanish educator and activist. Opened Spain's first modern school in Barcelona in 1902—secular, coeducational, open to rich and poor, and fiercely opposed by the church. Soon operated 40 such schools. Organized an International League of Rational Education. Accused of fomenting anticonscription and antireligious riots and strikes in 1909, and—following a military tribunal from which defense witnesses were excluded—executed. Pope Pius X sent the prosecutor a gold-handled sword engraved with his congratulations.*

> "The need for religion will end when man becomes sensible enough to govern himself."

From his will, written on his prison-cell wall on the eve of his execution:

> "Let no more gods or exploiters be served. Let us learn rather to love one another."

Ludwig von Feuerbach (1804–1872), *German materialist philosopher. His book* The Essence of Christianity *is a classic of atheist literature. A powerful influence on the Young Hegelians, including the young, barely bearded* KARL MARX.

> "It is clear as the sun and as evident as the day that there is no God, and still more that there can be none."

> "Atheism is the secret of religion. . . . Religion is nothing else than the consciousness of the infinity of the consciousness. . . . Divine revelation is simply the self-determination of man, only . . . between himself the determined, and himself the determining, he interposes an object—God. . . . God is the

medium by which man brings about the reconciliation of himself
with his own nature. . . . And so in revelation man goes out of
himself, in order, by a circuitous path, to return to himself!"

"To think is to be God." *Cogito ergo Deus sum?*

"The decline of culture was identical with the victory of
Christianity . . . religious man feels no need of culture."

"What yesterday was still religion is no longer such today; and
what today is atheism tomorrow will be religion."

Richard Feynman (1918–1988), *American Nobel-winning physicist/comedian/bongo player. Helped develop the atomic bomb. Did important work in particle theory, superconductivity, and quantum computing. First to publicly propose nanotechnology (1959). My college physics course consisted largely of his famous videotaped lectures. I thought he was some bozo. FREEMAN DYSON called him "half-genius, half-buffoon," but later changed this to "all-genius, all-buffoon." Liked to do some of his work in a topless bar. (Maybe that's where he first said, "Physics is to math what sex is to masturbation.") Last words: "I'd hate to die twice, it's so boring."*

"God was invented to explain mystery. . . . When you finally
discover how something works . . . you don't need him anymore.
But you need him for the other mysteries. So therefore you leave
him to create the universe because we haven't figured that out
yet . . . [and] to explain consciousness . . . stuff like that."

"I think it's much more interesting to live not knowing than to
have answers which might be wrong. . . . I don't feel frightened
by being lost in a mysterious universe without any purpose,
which is the way it really is as far as I can tell."

Charles Kellogg Field (1873–1948), *American poet and wit. San Francisco native. After the great S. F. earthquake/fire of 1906, clergymen said the catastrophe was divine retribution for the city's wicked ways (like JERRY FALWELL re 9/11). It so happened that a warehouse and its contents—thousands of barrels of highly flammable whisky—survived in the heart of the inferno. Field scribbled a ditty, later inscribed on a plaque that still hangs on the building:*

"If, as they say, God spanked this town / For being much too frisky, / Why did He burn / His churches down / And save Hotaling's Whiskey?"

Henry Fielding (1707–1754), *English novelist (Tom Jones, Joseph Andrews) and dramatist. One of his plays prompted the passage of a Theatrical Licensing Act that made political satire on the stage virtually impossible. As a justice of the peace, issued a warrant for the arrest of a playwright for "murder of the English language."*

"No man has ever sat down calmly unbiased to reason out his religion, and not ended by rejecting it."

"There are a set of religions, or rather moral writings, which teach that virtue is the certain road to happiness, and vice to misery, in this world. A very wholesome and comfortable doctrine, and to which we have but one objection, namely, that it is not true."

Emmett F. Fields, *Kentucky-based atheist, providing atheism services to the South and Midwest for over a quarter of a century. Operates the* Bank of Wisdom, *an online collection of writings and "suppressed books" on religion and philosophy.*

"The Atheist Bible, it could be said, has but one word: 'Think.'"

W. C. Fields (born William Claude Dukenfield, 1880–1946), *American comedian, drinker, and legendary misanthrope (legendary as in not true). Created such film and stage characters as Larson E. Whipsnade, Ambrose Wolfinger, T. Frothingill Bellows, and Professor Eustace P. McGargle. In his final weeks in a hospital, the known atheist was caught reading the Bible. Asked why, he explained, "I'm checking for loopholes." Died on Christmas Day, the holiday he said he hated.*

"[Prayers] may bring solace to the sap, the bigot, the ignorant, the aboriginal, and the lazy—but it is the same as asking Santa Claus to bring you something for Christmas."

Nothing to do with religion, but more immortal than God;—from Mississippi:

> "Whilst traveling through the Andes Mountains, we lost our corkscrew. Had to live on food and water for several days."

Harvey Fierstein (1954–) *gravelly-voiced, barrel-bodied, outrageously gay American actor/playwright. Won Tony awards for Best Actor and Best Play for* Torch Song Trilogy *(1983), for Best Book for the musical* La Cage Aux Folles *(1984), and for Best Actor as Tracy Turnblad's mountain-size mother Edna in* Hairspray.

> "I don't believe in God [or] in heaven or hell, but I pray three or four times a day. . . . [I pray] when I forget a line. You know how they say there are no atheists in the foxholes? Well, there's no atheist at the Minskoff [Theater] either."

> "The Catholic Church is the only organization on record to dispense money from a slush fund set up solely for the paying off of abused children's families. So always remember you cannot judge a man by his collar."

Geoffrey Fisher (1887–1972), *Archbishop of Canterbury (1945–1961). 1954:*

> "The hydrogen bomb is not the greatest danger of our time. After all, the most it could do would be to transfer vast numbers of human beings from this world to another and more vital one into which they would some day go anyway. . . ."[13,14]

F. (Francis) Scott Fitzgerald (1896–1940), *Irish-American novelist and short story writer. Married Zelda Sayre in New York's St. Patrick's Cathedral.*

> "You can take your choice between God and Sex. If you choose both you're a hypocrite; if neither, you get nothing."

Camille Flammarion (1842–1925), *French astronomer.*

"Men have had the vanity to pretend that the whole creation was made for them, while in reality the whole creation does not suspect their existence."

Gustave Flaubert (1821–1880), *French novelist. Spent five years writing* Madame Bovary, *whose realistic portrayal of adultery offended religious sensibilities and led to his criminal prosecution. He got off.*

"It is necessary to sleep upon the pillow of doubt."

"My kingdom is as wide as the universe and my wants have no limits. I go forward always, freeing spirits and weighing worlds, without fear, without compassion, without love, without God. I am called Science."

"Between the Immaculate Conception and free lunches for workingmen, everything marches toward ruin."

Antony Flew (1923–), *British philosopher and famous waffling atheist. Argued for decades that one should presuppose atheism until evidence of a God surfaces. Ugly rumors first surfaced in 2001 that he had decided such evidence exists. Flew refuted them on Secular Web, and in 2003, without torture or coercion, signed the Humanist Manifesto III. In a 2004 interview, however, he declared himself a deist who accepted God as a First Cause but rejected belief in an afterlife and the resurrection of Jesus. In fact, he okayed the title "Atheist Becomes Theist." The atheist world reeled. Backpedaling ensued. Then forepedaling. In May 2006 Flew accepted the "Phillip E. Johnson Award for Liberty and Truth"—named for one of the leading promoters of "intelligent design"—from a Christian university in California. His soul is in grave peril.*

The Flew that has flown: "Stuff is all there is; while everything which is not stuff is nonsense."

"You cannot transmute some incoherent mixture of words into sense merely by introducing the three-letter word 'God' to be its grammatical subject."

One Flew over the cuckoo's nest—early 2004: "Reason assures us that there is a God. . . . My one and only piece of relevant evidence is the apparent impossibility of providing a naturalistic theory of the origin from DNA of the first reproducing species."

December 2004: "I now realize that I have made a fool of myself by believing that there were no presentable theories of the development of inanimate matter up to the first living creature capable of reproduction. . . . I'm quite happy to believe in an inoffensive inactive god." *(At the same time, he denied having abandoned his atheism. Trial is set for March 12.)*

Larry Flynt (1942–), *American publisher of* Hustler *magazine. An assassination attempt in 1978 left Flynt paralyzed from the waist down and ended his one-year conversion to Christianity, which he subsequently chalked up to a manic-depressive episode. When it ended, he sought out the services of, and held a party for,* ST. MADALYN O'HAIR. *Ran for president as a Republican in 1984. Now a Dem. In 1988 the Supreme Court ruled against* JERRY FALWELL *in his lawsuit against Flynt over a satirical ad in* Hustler *that suggested Falwell's first sexual encounter was with his mother in an outhouse. Guess the Court believed it. His investigations of Republican lawmakers' sexual pecadillos during the Clinton impeachment proceedings led to the resignation of incoming House speaker Bob Livingston.*

"I have left my religious conversion behind and settled into a comfortable state of atheism. I have come to think that religion has caused more harm than any other idea since the beginning of time. The Jerry Falwells of this world are living proof of the hypocrisy that permeates organized religion in America and around the world."

G. W. (George William) Foote (1850–1915), *British atheist secularist activist. Prosecuted for blasphemy in 1882 and jailed for a year, but not in vain: The outcry by his supporters led to the legalization of criticism of Christianity. Coauthor of* The Bible Handbook *(1899), a freethought classic that methodically laid out (in the authors' words) the Good Book's "contradictions . . . absurdities . . . immoralities, indecencies, and obscenities."*

"The God who is to be the object of our adoration and imitation is depicted to us as a judge who will grant vengeance in answer to incessant prayer . . . as an employer who pays no more for a life-time than for the nominal service of a death-bed repentance."

"It will yet be the proud boast of women that they never contributed a line to the Bible." *Sexist bastard.*

Wil Forbis, *keeper of the "e-zine"* Acid Logic *and the blog* My So-called Penis: The Blog That Strokes Itself, *which devotes Fridays to dick-related news items.*

"I've yet to find anything convincing about the arguments Christians make for the existence of this God chap, and feel that if he does exist, events such as the Holocaust, Cambodian massacres and Limp Bizkit illuminate the fact that he's been asleep at the wheel for quite some time."

E. M. (Edward Morgan) Forster (1879–1970), *English novelist (*Howard's End, A Room with a View, A Passage to India, Where Angels Fear to Tread*). Prominent member of the British Humanist Association.*

"There lies at the back of every creed something terrible and hard for which the worshipper may one day be required to suffer."

"Faith, to my mind, is a stiffening process, a sort of mental starch, which ought to be applied as sparingly as possible. I dislike the stuff. . . . My motto is: 'Lord, I disbelieve—help thou my unbelief.'"

Anatole France (1844–1924), *French writer. His novel* The Revolt of the Angels *portrayed God as evil and the devil as benign.*

"If 50 million people believe a foolish thing, it is still a foolish thing."

"Nature has no [moral] principles . . . makes no distinction between good and evil."

"Religion has done love a great service by making it a sin." *A Frenchman can't be wrong!*

"You believe you are dying for the fatherland—you die for some industrialists."

Al Franken (1951–), *American comedian and liberal political commentator.* Saturday Night Live *writer/cast member for 15 years. Air America radio host. (Show's original name:* The O'Franken Factor.*) Author of five #1* New York Times *bestsellers. Graduated cum laude from Harvard with a BA in government.*

"[Members of the religious right] sometimes forget we don't live in a theocracy. They can be in the public square and express their opinion but to expect other people to alter their behavior to say that, for example, that homosexuality is immoral because it says so in the Bible. . . . I mean it also says you can't eat pork. I don't see a lot of orthodox Jews saying people who eat pork shouldn't be allowed to get insurance benefits. I mean there's stuff in the bible how about how to sell your daughter."

On JERRY FALWELL *saying the Anti-Christ is a living Jewish male but he didn't know whether it was Marvin Hamlisch:*

"I thought you could rule out Hamlisch. . . . Why would the Anti-Christ write *Chorus Line?*"

Benjamin Franklin (1706–1790), *American Founding Father, inventor and, like* JEFFERSON, ADAMS, *and other F.F.s, a Deist.*

"Lighthouses are more helpful than churches."

"The way to see by faith is to shut the eye of reason."

"Since it is impossible for me to have any positive, clear idea of that which is infinite and incomprehensible, I cannot conceive otherwise than that He . . . expects or requires no worship or praise from us."

"As to Jesus of Nazareth . . . I have some doubts as to his divinity; though it is a question I think it needless to busy myself with now, when I expect soon an opportunity of knowing the truth with less trouble."

"Indeed, when religious people quarrel about religion, or hungry people quarrel about victuals, it looks as if they had not much of either among them."

Sir James Frazer (1854–1941), *Scottish anthropologist. When first published, his book* The Golden Bough, *a huge comparative study of religion and mythology—the work of 25 years—scandalized the public because it argued that almost all the world's religions and mythologies— including the Christian story of Jesus—are variations on the same theme, which grows out of ancient fertility cults and centers on the sacrifice of a sacred king, the incarnation of a dying and reviving god. Frazer removed his analysis of the Crucifixion from later editions.*

"Men make the gods; women worship them."

"[T]he fear of the human dead . . . I believe to have been probably the most powerful force in the making of primitive religion."

Frederick the Great (1712–1786), *Prussian king. Zealous defender of Enlightenment ideas. Patron of* VOLTAIRE. *. . . . to whom he remarked:*

"There are so many things to be said against religion that I wonder they do not occur to everyone."

Fred's minister of religion, replying to an assertion by church elders that "those who believe most are the best subjects":

"His Majesty is not disposed to rest the security of his state upon the stupidity of his subjects."

Timothy Freke, *British philosopher and licensed mystic. Author of some two-dozen books on religion and mysticism. He and Peter Gandy*

coauthored the top-ten bestseller The Jesus Mysteries: Was the Original Jesus a Pagan God? *(yes) and* Jesus and the Lost Goddess *(2002), cited by Dan Brown as an inspiration for* The Da Vinci Code.

> "The great irony is that, if they could but see it, Christian and Islamic Fundamentalists are the same people. Their vision of life and how to live it is driven by the same needs and neuroses. What they hate in each other is a projection of what they hate in themselves. . . . It is Literalists who fight wars of religion with Literalists from other traditions, each claiming that God is on their side."

> "If Jesus is the one and only Son of God who requires the faithful to acknowledge this as historical fact, then Christianity must be in opposition to all other religions who do not teach this. Moreover, if all unbelievers are to be damned for eternity it becomes the moral duty of Literalist Christians to spread their beliefs, by force if necessary, to save as many souls as possible, even if it means destroying their bodies to do so."

Sigmund Freud (1856–1939), *Austrian-Jewish physician and psychoanalyst. Raised in a nonreligious home by a freethinking father. To Freud, religion was part of humankind's infancy, an outgrowth of psychological needs. "A personal god was nothing more than an exalted father-figure: desire for such a deity sprang from infantile yearnings for a powerful, protective father, for justice and fairness and for life to go on forever," wrote* KAREN ARMSTRONG. *"Religion may have been the original cure," wrote biographer Philip Rieff; "Freud reminds us that it was also the original disease."*

> "Religion is the universal obsessional neurosis of mankind; like the obsessional neurosis of children, it arose out of the Oedipus complex, out of the relation to the father. . . . [It is] a parallel to the neurosis which the civilized individual must pass through on his way from childhood to maturity."

> "The God-Creator is openly called Father. Psychoanalysis concludes that he really is the father, clothed in the grandeur in which he once appeared to the small child. . . . The emotional strength of this memory-image and the lasting nature of his need for protection are the two supports for [the religious man's] belief in God."

"Devout believers are safeguarded in a high degree against the risk of certain neurotic illnesses; their acceptance of the universal neurosis spares them the task of constructing the personal one."

"Think of the depressing contrast between the radiant intelligence of a healthy child and the feeble intellectual powers of the average adult. Can we be quite certain that it is not precisely religious education which bears a large share of the blame for this relative atrophy?"

Like other neuroses or addictions, religion can be overcome, but only by facing up to the truth: "They will have to [*Step One*] admit to themselves the full extent of their helplessness and their insignificance in the machinery of the universe; they can no longer be the centre of creation, no longer the object of tender care on the part of a beneficent Providence. . . . We may call this 'education to reality.'. . . It is something, at any rate, to know that one is thrown upon one's own resources. One learns then to make a proper use of them."

Erich Fromm (1900–1980), *German-Jewish-American psychologist/philosopher associated with the left-wing Frankfurt School of critical thinkers (also see HERBERT MARCUSE). One of the founders of socialist humanism. Author of popular books combining psychology, philosophy, and social and political commentary. His grandfather and two great grandfathers were rabbis. His own study of Talmud became central to his worldview, even though he turned away from orthodox Judaism and toward secular interpretations of scripture.*

"Once a doctrine, however irrational, has gained power in a society, millions of people will believe it rather than feel ostracised and isolated."

"Theologians and philosophers have been saying for a century that God is dead, but what we confront is the possibility that man is dead, transformed into a thing, a producer, a consumer, and idolator . . ."

Robert Frost (1874–1963), *American poet. Raised by his mother in the Swedenborgian church (founded on the belief that Emanuel Swedenborg,*

an eighteenth-century scientist and mystic, witnessed the Last Judgment and the Second Coming and that his writings form a third testament of the Bible), but escaped from it as an adult. Winner of four Pulitzer Prizes.

"I turned to speak to God / About the world's despair; / But to make bad matters worse, / I found God wasn't there." *You know, poetry doesn't look that hard.*

"I hold it to be the inalienable right of anybody to go to hell in his own way."

"Don't be an agnostic—be something."

Charles E. Fuller *(1887–1968), American Baptist clergyman and popular radio evangelist.*

"Fellowship with God means warfare with the world." *We've noticed.*

R. Buckminster Fuller *(1895–1983), American engineer/inventor/philosopher/visionary/guru. Invented the geodesic dome, Dymaxion car, Dymaxion home. Apostle of solar and wind power, recycling, efficiency, tetrahedral geometry, and unprecedented universal wealth. Everything had to be rethought from scratch: buildings, cars, boats, bathrooms, maps, language. (Coined the terms synergetics, tensegrity, ephemeralization. . . . How did we live without those?) Documented his life every 15 minutes from 1915 to 1983, making his the most documented human life in history. A Unitarian-Universalist. (Barely a religion; more like "be nice.")*

"God, to me, is a verb."

"Sometimes I think we're alone. Sometimes I think we're not. In either case, the thought is quite staggering."

Fundamentalists Anonymous*: One of several support groups for "ex-tians" and other recovering Godoholics. Guess how many steps their program has? The full complement. (Religion abhors a vaccum. . . .)*

1. "I realize that I had turned control of my mind over to another person or group, who had assumed power over my thinking." *(Plus 11 more.)*

Robert W. Funk (1926–2005), *Bible scholar. Chairman of the graduate department of religion, Vanderbilt University. Founder in 1985 of the Jesus Seminar, a controversial research team of about 100 scholars devoted to scientifically reconstructing the life of Jesus. Author of* Honest to Jesus, A Credible Jesus, The Acts of Jesus: The Search for the Authentic Deed, *and* Turning Jesus Upside-Down and Shaking Him Until All the Change Falls Out of His Pockets. *No, not that last one.*

"If the evidence supports the historical accuracy of the gospels, where is the need for faith? And if the historical reliability of the gospels is so obvious, why have so many scholars failed to appreciate the incontestable nature of the evidence?"

SECULAR * INFIDEL * NONBELIEVER * HUMANIST * RATIONALIST * FREETHINKER * AGNOSTIC * GODLESS * HERETIC * ATHEIST *

G

Yuri Gagarin (1934–1968), *Soviet-Russian cosmonaut; first atheist (and first human) in space (1961).*

"I don't see any god up here."

Matilda Joslyn Gage (1826–1898), *another of this book's 288 nineteenth-century American suffragist-abolitionist-freethinker-writer-editor-activists, each one of whom seems to have founded the most important women's rights organization in history, been the daughter of prominent New England abolitionists, and had three names. Wrote* Woman, Church and State, *about Christianity's role in the oppression of women. Coauthored* ELIZABETH CADY STANTON's *heretically feminist* Woman's Bible. *President of the National Woman Suffrage Association. Mother-in-law of* Wizard of Oz *author L. Frank Baum. The tendency of woman scientists to receive less credit than their work merits has been named, after her, the Matilda Effect.*

"Believing this country to be a political and not a religious organisation . . . the editor of the *National Citizen* will use all her influence of voice and pen against 'Sabbath Laws,' the uses of the 'Bible in School,' and pre-eminently against an amendment which shall introduce 'God in the Constitution.'"

Galileo Galilei (1564–1642) *Italian physicist, astronomer, and philosopher. A devout Catholic, let's not kid ourselves; but for rejecting blind allegiance to authority, whether of the church or of Aristotle, he is called the "father of modern science" and treated by atheist as a saint. Discovered the first and second laws of motion. Made the first microscope. His astronomical observations confirmed that Copernicus revolved around the sun—and got Galileo busted for heresy by the Holy Office, a.k.a. Inquisition, which forced him to recant.*

Said to have murmured right afterwards: "Epur si muove"—It [the Earth] "still does move."

"I do not feel obliged to believe that the same God who has endowed us with sense, reason, and intellect has intended us to forgo their use. . . . In the discussion of natural problems we ought to begin not with the Scriptures, but with experiments, and demonstrations."

Liam Gallagher (1972–), *lead singer with the British band Oasis. The tabloid press feasted, nay, gorged on stories of his alleged drug use, sexual promiscuity, fights, and all-around rock'n'roll bad-boy-ness. Banned for life from Cathay Pacific airlines after a row over a scone. Obsessed with JOHN LENNON, of whom he has claimed to be the reincarnation (which would mean there were two John Lennons for eight years—but part of this book's message is that there's more to life than rationalism).*

"If I die and there's something afterwards, I'm going to hell, not heaven. I mean, the devil's got all the good gear. What's God got? The [rival British band] Inspiral Carpets and nuns. Fuck that."

How's this for atheism? "I respect the Stones but their songs are a pile of crap."

Mohandas "Mahatma" Gandhi (1869–1948), *alarmingly thin Indian political and spiritual leader. Pioneer of civil disobedience, which helped lead India to independence. Derived most of his principles from*

his Hindu faith, in which, oddly enough, he only became deeply interested after meeting English students of Indian religion at the British Vegetarian Society, which he joined while a student in London to help keep his vow to his devout mother not to eat meat. Blamed a terrible earthquake in India in 1934 on the sin committed by upper caste Hindus by not letting untouchables into their temples. Murdered by a Hindu extremist who thought Gandhi was appeasing Muslim Pakistan. Mahatma means Great Soul.

Asked if he was a Hindu: "Yes I am. I am also a Christian, a Muslim, a Buddhist and a Jew."

"God has no religion."

"I like your Christ, I do not like your Christians. Your Christians are so unlike your Christ."

"An eye for an eye makes the whole world blind."

"God is conscience. He is even the atheism of the atheist." *Then consciousness of godlessness is God? My head is spinning . . . I'm losing consciousness. . . .*

"If you don't find God in the next person you meet, it is a waste of time looking for him further."

Helen Hamilton Gardener (1853–1925), *American suffragist / feminist activist. Appointed in 1920 to the Civil Service Commission—the highest office a woman had yet occupied in the federal government. Author of* Men, Women and Gods. *Dubbed by the* New York Sun *"[RALPH] INGERSOLL done in soprano," and by the* Chicago Times, *"the pretty infidel." Patronizing, sexist pigs.*

"I do not know the needs of a god or of another world. . . . I do know that women make shirts for seventy cents a dozen in this one."

Théophile Gautier (1811–1872), *French poet-dramatist-critic. Popularized the slogan "l'art pour l'art," "art for art's sake."*

"Virginity, mysticism, melancholy! Three unknown words, three new maladies brought by Christ."

Bob Geldof (1951–), *Irish rock musician, humanitarian and Sir. Asked in 2006, Are you a saint or a sinner?*

"Being an atheist I can't be either."

Edward Gibbon (1737–1794), *English historian. Author of* The Decline and Fall of the Roman Empire, *now available in a handy 1,900-page edition.*

"The evidence of the heavenly witnesses—the Father, the Word, and the Holy Ghost—would now be rejected in any court of justice."

"So urgent on the vulgar is the necessity of believing, that the fall of any system of mythology will most probably be succeeded by the introduction of some other mode of superstition."

André Gide (1869–1951), *French author. Nobel Prize for Literature, 1947. In his work, Gide "exposes to public view the conflict and eventual reconciliation between the two sides of his personality . . . as he perceives himself: the austere and refined Protestant, and the divinely inspired—and no longer blushing—pederast."[1] Discovered that "orientation"—and befriended* OSCAR WILDE—*on a sojourn in Algiers in 1893–1894. Had a 16-year-old male lover when he was 47, while in an unconsummated marriage. As though trying to improve his sales, the Catholic Church placed his works on the Index of Forbidden Books in 1952.*

"Christianity, above all, consoles; but there are naturally happy souls who do not need consolation. Consequently, Christianity begins by making such souls unhappy, for otherwise it would have no power over them."

"It is much more difficult than one thinks not to believe in God." *So if at first you don't succeed. . . .*

Terry Gilliam (1940–), *American-born British film director/writer/ animator; member of the comedy group Monty Python. The* Onion

headline *"Terry Gilliam Barbecue Plagued by Production Delays"* adequately describes the fortunes of many of his films, which tend to be extremely expensive. Codirected Monty Python and the Holy Grail *with Python* TERRY JONES. *Working titles for the Python film* Life of Brian *included the infinitely funnier* Brian of Nazareth *and* Jesus Christ: Lust for Glory. *The film was banned in Ireland, Norway, and some areas of England, owing above all to the final scene in which a chorus of crucified men, led by Brian, sings "Always Look at the Bright Side of Life."*

> *Oh, lighten up:* "With *Life of Brian*, we were vilified by Christians. Yet Christianity is alive and well. [*So the movie failed. At least they tried.*] Come on, if your religion is so vulnerable that a little bit of disrespect is going to bring it down, it's not worth believing in, frankly."

Tom Gilroy, *American actor/playwright/director.*

> "'Being Christian' is no longer defined by doing good deeds [but] by an arrogant mission to tell others how they must live—who they can marry, who they can adopt, what they must teach in schools. . . . Our national conversation on ethics, morality, and faith has become a kind of WWF [World Wrestling Federation] 'Religious Smackdown.' . . . But [the Bushies have] done us an odd—if unintentional—service by showing us in practice exactly what the Founding Fathers feared and tried to prevent."

Nikki Giovanni (Yolande Cornelia Giovanni, 1943–), *African-Italian-American poet, author, and professor of writing and literature at Virginia Polytechnic.*

> "White people really deal more with God and black people with Jesus."

Johann Wolfgang von Goethe (1749–1832), *the Goliath, titan, and colossus of German literature. Accused by the clergy of— (better sit down)—atheism. Sometimes referred to himself as a heathen or pagan. Spoke disdainfully of "the fairy-tale of Christ." "Goethe said those*

who have science and art have religion; and added, let those who have not science and art have the popular faith; let them have this escape, because the others are closed to them."—JOHN BURROUGHS.

"[SPINOZA] does not have to prove the existence of God. Being is God. If others denounce him as an atheist for this, I wish to exalt him."

"I would be well pleased if after the close of this life we should be blessed with another, but I would beg not to have there for companions any who have believed it here."

Emma Goldman *(1869–1940), Russian-Jewish-American socialist/anarchist activist and writer. President Herbert Hoover had her deported to the Soviet Union when she died, her body was brought back to the U.S. and buried in Chicago, beside the graves of fellow Chicago radicals, including VOLTAIRINE DE CLEYRE, whose biography she had written. From her 1916 essay "The Philosophy of Atheism":*

"The God idea is growing more impersonal and nebulous in proportion as the human mind is learning to understand natural phenomena and [as] science progressively correlates human and social events. . . . God, today, no longer directs human destiny with the same iron hand as of yore. Rather does the God idea express a sort of spiritualistic stimulus to satisfy the fads and fancies of every shade of human weakness."

"Religion, 'Divine Truth,' rewards and punishments are the trademarks of the largest, the most corrupt and pernicious, the most powerful and lucrative industry in the world. . . . [But the theists] realize that the masses are growing daily more atheistic, more anti-religious . . . [more] engrossed in the problems of their immediate existence. . . . How to bring the masses back to the God idea—that is the most pressing question to all theists. . . . All these frantic efforts find approval and support from the earthly powers; from the Russian despot to the American president; from Rockefeller and Wannamaker down. . . . They know that capital invested in BILLY SUNDAY [*the leading evangelist of the day*], the YMCA, Christian Science, and various other religious institutions will return enormous profits from the subdued, the tamed, and dull masses. . . . They know that

Christianity is a more powerful protection against rebellion and discontent than the club or the gun."

"It is characteristic of theistic 'tolerance' that no one really cares what the people believe in, just so they believe or pretend to believe."

Barry Goldwater (1909–1998), *five-term Republican U.S. senator from Arizona whose 1964 presidential candidacy started the conservative resurgence—even though he got shmeared, as his Jewish-born father might have said. Accomplished amateur photographer, ham radio operator, drinker (not so amateur), and UFO buff who believed the U.S. government was withholding UFO evidence: "I certainly believe in aliens in space," he told* LARRY KING.

"There is no more powerful ally one can claim in a debate than Jesus Christ, or God, or Allah. . . . But like any powerful weapon, the use of God's name on one's behalf should be used sparingly. . . . I'm frankly sick and tired of the political preachers across this country telling me as a citizen that if I want to be a moral person, I must believe in A, B, C, and D. Just who do they think they are? . . . I am even more angry as a legislator who must endure the threats of every religious group who thinks it has some God-granted right to control my vote on every roll call in the Senate." *This from the founder of the modern conservative movement! (And in 1981! In case you thought this rupture in the Sacred Wall of Separation began with Bush.)*

Edmond de Goncourt (1822–1896), *French writer, critic, and publisher. The annual Prix Goncourt, which he in his a-religious wisdom endowed, is the most prestigious French literary prize.*

"If there is a God, atheism must seem to Him as less of an insult than religion." *(The insult of presuming on His attention and expecting personalized services, presumably.)*

Gora (Goparaju Ramachandra Rao, 1902–1975), *Indian atheist leader. In 1940 he and his wife founded India's Atheist Center, which has campaigned to abolish the caste system and child marriages and to rid India of belief in karma or divine fate, which only reconciles the poor to their suffering. The center received the International Humanist and Ethical Union's International Humanist Award in 1986. Gora, a confidante of* GANDHI, *married his wife when she was 10—normal in 1922 (Orthodox Hinduism dictates that girls must marry before puberty), illegal today, thanks only to secularists' efforts.*

> "Because morality is a social necessity, the moment faith in god is banished, man's gaze turns from god to man and he becomes socially conscious. Religious belief prevented the growth of a sense of realism. But atheism at once makes man realistic and alive to the needs of morality."

Lydia Gottschewski, *German Nazi writer and organizer of the League of German Girls under* HITLER. *(Quoted simply as a "German political activist" on at least one atheist Web site and on Feminist.com's "Women of Wisdom" page—"In sisterhood, Elaine Bernstein Partnow, Editor.") Commenting on another of religion's virtues in her book* Women in the New State, *1934:*

> "It is a curious fact that pacifism . . . is a mark of an age weak in faith, whereas the people of religious times have honored war as God's rod of chastisement. . . . Only the age of enlightenment has wished to decide the great questions of world history at the table of diplomats." *Couldn't have put it better ourselves.*

Stephen Jay Gould (1941–2002) *American paleontologist, evolutionary biologist, and popular science writer. Coined the term "Non-Overlapping Magistèria" (NOMA) for his view that science and religion do not conflict—they occupy different realms and can have nothing to say about one another's claims. This is hotly denied by folks like* DAWKINS *and* DENNETT. *Once voiced a cartoon version of himself on* The Simpsons.

"If you absolutely forced me to bet on the existence of a conventional anthropomorphic deity, of course I'd bet no. But, basically, [THOMAS] HUXLEY was right when he said that agnosticism is the only honorable position because we really cannot know."

"We are here because one odd group of fishes had a peculiar fin anatomy that could transform into legs for terrestrial creatures; because the earth never froze entirely during an ice age; because a small and tenuous species, arising in Africa a quarter of a million years ago, has managed, so far, to survive by hook and by crook. We may yearn for a higher answer—but none exists."

Remy de Gourmont (1858–1915), *French Symbolist writer, critic, and philosopher. Influenced EZRA POUND and T. S. Eliot. Dismissed from his librarian position in the Bibliothèque Nationale in Paris for mocking French patriotism. (Patriotism: the last refuge of religious zeal, when God has been discovered missing?)*

"God is not all that exists. God is all that does not exist."

"Religions revolve madly around sexual questions."

Robert Graves (1895–1985), *British novelist, poet, and classical scholar. His works include the historical novels* I, Claudius *and* King Jesus. The White Goddess *(1948), his study of goddess-worship as the proto-typical religion/mythology of Europe, was inspired by JAMES FRAZER's* Golden Bough. *Graves's book fueled the post-1960s raft of feminist writings in celebration of ancient, pre-"patriarchal" goddess worship and the kinder, gentler, matriarchal cultures it encouraged. You see, we men deposed the goddess, set up repressive, warmongering male gods like Yahweh in Her place, and the whole world's been fucked up ever since. (Graves's own scholarship was pretty flaky. His goddess is said to bear a strong resemblance to his longtime lover and muse, Laura Riding.)*

"By Jesus's time the Law of Moses, originally established for the government of a semi-barbarous nation of herdsmen and hill-farmers, resembled a petulant great-grandfather who tries to govern a family business from his sick-bed . . . unaware of the changes that have taken place in the world since he was able to get about. . . ."

"What the scientist thinks today, everyone else will be thinking on the day after tomorrow."

Ruth Hurmence Green (1915–1981), *American author of The Born-Again Skeptic's Guide to the Bible. Became an atheist after surviving cancer. Another God-free foxhole, as it were. (Name your baby Godfree. If it's a girl, Atheista or Secularia.)*

"It is possible to pull out justification for imposing your will on others, simply by calling your will God's will."

"If the concept of a father who plots to have his own son put to death is presented to children as beautiful and as worthy of society's admiration, what types of human behavior can be presented to them as reprehensible?"

Pope Gregory I ("the Great") (540–604). *Made his name by getting the Patriarch of Constantinople burned at the stake for writing that the resurrection of the dead would be incorporeal. Well, I say the Patriarch had it coming.*

"The bliss of the elect in heaven would not be perfect unless they were able to look across the abyss and enjoy the agonies of their brethren in eternal fire."

Pope Gregory VI (d. 1048). *Purchased the papacy from his 20-year-old godson, Pope Benedict IX, who wanted to abdicate so he could marry.*

"From the polluted fountain [of] that absurd and erroneous doctrine, or rather raving, which claims and defends liberty of conscience for everyone . . . comes, in a word, the worst plague of all—liberty of opinions and free speech."

Matt Groening (1954–), *American cartoonist: Creator of the animated TV series* The Simpsons *and* Futurama *and of the comic strip* Life in Hell, *loosely inspired, believe it or not, by a chapter titled "How to Go to Hell" in* WALTER KAUFMANN'S *book* Critique of Religion and Philosophy. *The strip spun off the books* School Is Hell, Childhood Is Hell, Work is Hell (amen), The Big Book of Hell, *and* The Huge Book of Hell.

Asked what he considers the most comical story in the Bible: "I was very disturbed when Jesus found a demon in a guy, and he put the demon in a herd of pigs, then sent them off a cliff [Mark 5.12–13]. What did the pigs do? I could never figure that out. It just seemed very un-Christian."

Bart Simpson: "Dear God, we paid for all this stuff ourselves, so thanks for nothing."

The Simpsons' *Rev. Lovejoy, in a sermon about the "Movementarians":* "This so-called new religion is nothing but a pack of weird rituals and chants designed to take away the money of fools. Let us say the Lord's prayer 40 times, but first let's pass the collection plate. . . . And as we pass the collection plate, please give as if the person next to you was watching."

Superintendent Chalmers: "A prayer in a public school! God has no place within these walls, just like facts have no place within organized religion."

Alan Guth (1947–), *American physicist and cosmologist at MIT. Father of "cosmic inflation"—the idea that the nascent universe passed through a phase of exponential expansion driven, obviously, by a negative vacuum energy density, and that the cost of living is increasing exponentially throughout the universe, resulting in negative vaccum cash density.*

"The universe could have evolved from absolutely nothing in a manner consistent with all known conservation laws. . . . The question of the origin of the matter in the universe is no longer thought to be beyond the range of science . . . *everything* can be created from nothing. . . . It is fair to say that the universe is the ultimate free lunch."

Jetsun Jamphel Ngawang Lobsang Yeshe Tenzin Gyatso, a.k.a. The (14th) Dalai Lama ("Holy Lord, Gentle Glory, Compassionate, Defender of the Faith, Ocean of Wisdom") (Born Lhamo Thondup, 1935–) *Tibetan spiritual leader. As per tradition, a search party found and identified Thondup at around age three as the new incarnation of the DL. Presented with relics and toys that had belonged to the previous DL along with some that hadn't, little Thondup correctly identified all those that had by crying: "It's mine! It's mine!" It was so cute. . . . Until you've had a little lama of your own, you just can't understand.*

"According to Buddhism, one's own actions are the creator, ultimately. . . . From a certain angle, Buddhism is not a religion but rather a science of mind. . . . We must conduct research and then accept the results. If they don't stand up to experimentation, BUDDHA's own words must be rejected."

SECULAR * INFIDEL * NONBELIEVER * HUMANIST * RATIONALIST * FREETHINKER * AGNOSTIC * GODLESS * HERETIC * ATHEIST *

Ernst Haeckel (1834–1919), *German evolutionary biologist/ philosopher. Tried to found an organized pantheist religion. "Yet his misguided interpretation of Darwinism led him to a brutal social ethic which influenced and gave spurious scientific legitimacy to the Nazi programme."[7] Worse, Haeckel's idea and phrase "ontogeny recapitulates phylogeny"—i.e., the developing embryo diplays the evolutionary stages of its species—became a popular device for showing off intellect at parties. When* The Origin of Species *was published, Haeckel predicted that fossils of an ancestral human would be found in Indonesia, named the species, and instructed his students to go find it. One student went, and dug up "Java man," a primitve tool maker and coffee drinker and the first human ancestral remains ever found.*

> "God is everywhere identical with nature itself, and is operative within the world as force or energy. . . . Pantheism is the world system of the modern scientist. . . . The paths which lead to the noble divinity of truth and knowledge are the loving study of nature and its laws . . . not senseless ceremonies and unthinking prayers."

> "Atheism . . . is only another expression for [pantheism], emphasizing its negative aspect, the non-existence of any supernatural deity. In this sense SCHOPENHAUER justly remarks: 'Pantheism is only a polite form of atheism.'"

J. B. S. Haldane (1892–1964), *British geneticist, evolutionary biologist, and science popularizer. One of the founders of population genetics and developers of the mathematical theory of natural selection.*

"I believe that the scientist is trying to express absolute truth and the artist absolute beauty, so that I find in science and art, and in an attempt to lead a good life, all the religion that I want."

"My own suspicion is that the Universe is not only queerer than we suppose, but queerer than we can suppose."

E. Haldeman-Julius (1889–1951), *American publisher and editor of a muckraking socialist newspaper,* Appeal to Reason. *Author of* The Meaning of Atheism.

"The influences that have lifted the race to a higher moral level are education, freedom, leisure, the humanizing tendency of a better-supplied and more interesting life. In a word, science and liberalism . . . have accomplished the very things for which religion claims the credit."

Butch Hancock (1945–), *American folk-country-rock singer/ songwriter; member of the group The Flatlanders. Described by* Rolling Stone *as "a raspy-voiced West Texas mystic with an equal affinity for romantic border balladry and Zen paradox."*

"Life in Lubbock, Texas taught me two things. One is that God loves you and you're going to burn in hell. The other is that sex is the most awful, dirty thing on the face of the earth and you should save it for someone you love."

Jack Handey (1949–), *American humorist. Two sardonic Texans in a row! Best known for Deep Thoughts, surrealistic musings featured on* Saturday Night Live *and collected in several volumes.*

"We tend to scoff at the beliefs of the ancients . . . but we can't scoff at them personally, to their faces, and this is what really annoys me."

"If a kid asks where rain comes from, I think a cute thing to tell him is 'God is crying.' And if he asks why God is crying, another cute thing to tell him is, 'Probably because of something you did.'"

"If God dwells inside us like some people say, I sure hope He likes enchiladas, because that's what He's getting."

"I wish I had a kryptonite cross, because then you could keep both Dracula and Superman away."

Rt. Rev. Richard Harries (1936–), *English clergyman; Bishop of Oxford. Declared in May 2006 that homosexual unions are supported by the Bible, gays should be allowed to become bishops, and traditionalists need to be "converted." (To tolerance, not to homosexuality. Not that there would be anything wrong with that.)*

"Historians of science note how quickly the late Victorian Christian public accepted evolution. It is therefore quite extraordinary that 140 years later [2002], after so much evidence has accumulated, a [state-funded] school in Gateshead [U.K.] is opposing evolutionary theory on alleged biblical grounds. Do some people really think that the worldwide scientific community is engaged in a massive conspiracy to hoodwink the rest of us?"

Michael Harrington (1928–1989), *America's best-known socialist during his lifetime. Headed Democratic Socialists of America, whose members have included John Sweeney, Irving Howe, GLORIA STEINEM, Ed Asner, BARBARA EHRENREICH, Cornel West, and NOAM CHOMSKY. Jesuit-educated; one-time member of the Catholic Worker movement.*

"It is relevant to my present attitudes that even though I rejected the Church . . . I clearly remain a 'cultural Catholic,' much as an atheist Jew is culturally Jewish. . . . To complicate matters further, I consider myself to be—in [sociologist] Max Weber's phrase— 'religiously musical' even though I do not believe in God. . . . I am, then, what [sociologist] Georg Simmel called a 'religious nature without religion,' a pious man of deep faith, but not in the supernatural."

Sam Harris (1967–) *American author of the bestselling 2005 book*
The End of Faith: Religion, Terror, and the Future of Reason—*the second-best book on religion on the market today. The threat the world faces, it argues, is not religious extremism—it is religion.*

"Tell a devout Christian that his wife is cheating on him, or that frozen yogurt can make a man invisible, and he is likely to require as much evidence as anyone else, and to be persuaded only to the extent that you give it. Tell him that the book he keeps by his bed was written by an invisible deity who will punish him with fire for eternity if he fails to accept its every incredible claim about the universe, and he seems to require no evidence whatsoever."

"No one is ever faulted in our culture for not 'respecting' another person's beliefs about mathematics or history. . . . When people make outlandish claims, without evidence, we stop listening to them—except on matters of faith."

"How comforting would it be to hear the President of the United States assure us that almighty Zeus is on our side in our war on terrorism? The mere change of a single word in his speech—from God to Zeus—would precipitate a national emergency. If I believe that Christ was born of a virgin, resurrected bodily after death, and is now literally transformed into a wafer at the Mass, I can still function as a respected member of society . . . because millions of others believe [the same]. . . . The perversity of religion is that it allows sane people to believe the unbelievable *en masse*."

"Anyone who thinks western or Israeli imperialism solves the riddle of Muslim violence must explain why we don't see Tibetan suicide bombers killing Chinese children. The Tibetans have suffered every bit as much as the Palestinians. . . . Where are the throngs of Tibetans seething with hatred, calling for the deaths of the Chinese? . . . What is the difference that makes the difference? Religion. . . . Read the Koran. Osama bin Laden is playing it more or less by the book. Anyone who says that there is no basis for his worldview in the doctrine of Islam is either dangerously ignorant or just dangerous."

"We can no longer ignore the fact that billions of our neighbors believe in the metaphysics of martyrdom, or in the literal truth of the book of Revelation, or any of the other fantastical notions that have lurked in the minds of the faithful for millennia—because our neighbors are now armed with chemical, biological, and nuclear weapons. . . . Words like 'God' and 'Allah' must go the way of 'Apollo' and 'Baal,' or they will unmake our world."

Hubert Harrison (1883–1927), *African-American writer. Central figure in the 1920s Harlem Renaissance. Nicknamed the Black Socrates. Once said everyone knew about the many errors in the Bible, "except perhaps in America." Plus ça change. Lamented that African Americans with agnostic views were rare, and "these are seldom, if ever, openly avowed. . . . I am inclined to believe that freedom of thought must come from freedom of circumstance."*

"It should seem that Negroes, of all Americans, would be found in the Free-thought fold, since they have suffered more than any other class of Americans from the dubious blessings of Christianity. . . . The church saw to it that the religion taught to slaves should stress the servile virtues of subservience and content. . . . It was the Bible that constituted the divine sanction of this 'peculiar institution.'"

"The power of [Jesus'] personality haunted me for a long time, but in the end that also went. Now I am an agnostic; not a dogmatic disbeliever nor a bumptious and narrow infidel . . . not at all of Col. [RALPH] INGERSOLL's school [but rather] such an agnostic as [THOMAS] HUXLEY was."

"I wish to admit here something that most Agnostics are unwilling to admit. . . . that Reason alone has failed to satisfy all my needs. For there are needs, not merely ethical but spiritual, inspirational . . . and these also must be filled."

Bret Harte (1836–1902), *American author. Best known for his first-hand accounts of pioneering life in California.*

"The creator who could put a cancer in a believer's stomach is above being interfered with by prayers."

John Hattan, *cochair (with his wife Shelley) of Metroplex (Dallas/Fort Worth) Atheists,* and *Grand High UberPope of the First Church of Shatnerology (FCS—full name: The Most Holy-n-High Church of the Blinding Light of the Holy Glowing™ Form of the One Toupeed and Gloriously Bloated Shatner). "Here at the [FCS], we worship the holy essences of the most benevolent ShatnerBeing! We are transfixed by his magnificent TOUPEE and girth! As you read, you will learn secrets that will change your life!"*

> "I understand prayer quite well. It's a masturbatory exercise that gives catharsis to the pray-er and a placebo effect to the pray-ee, but only if the pray-ee knows he's being prayed for. . . . Why not masturbate instead? It's basically the same as praying. It makes you feel good, it doesn't do any harm, it doesn't do any good, you can do it alone, or you can do it in groups!"

Stephen Hawking (1942–), *British theoretical physicist. Has made important discoveries concerning the Big Bang and black holes. ("A black hole is where God is dividing by zero," goes a centuries old physics joke.) Almost completely paralyzed by Lou Gehrig's disease, first diagnosed in 1963. The doctors gave him no more than three years to live. Has kept his outmoded, 1980s-era voice synthesizer (which has an American accent) because he identifies with it. Has explained that the concept of time has no meaning before the beginning of the universe; the question "What came before the Big Bang?" makes no more sense than "What lies north of the north pole?" So, could God, as it were, lie north of the north pole? More importantly—could Santa? The Catholic Church seized on the Big Bang theory when it was first proposed, officially declaring it to be in accordance with the Bible. The universe had a beginning—a Creation! Gloria in Excelsis Deo? Not so fast, Your Eminences and Emptinesses.*

> "What I have done is to show that it is possible for the way the universe began to be determined by the laws of science. In that case, it would not be necessary to appeal to God to decide how the universe began. This doesn't prove that there is no God, only that God is not necessary. . . . One does not have to appeal to God to set the initial conditions for the creation of the universe, but if one does He would have to act through the laws of physics." *That's right; He's not above the law.*

"The quantum theory of gravity has opened up a new possibility, in which there would be no boundary to space-time. . . . The universe would be completely self-contained and not affected by anything outside itself. It would neither be created nor destroyed. It would just *be*. . . . [This] also has profound implications for the role of God in the affairs of the universe. . . . So long as the universe had a beginning, we could suppose it had a creator. But if the universe is really completely self-contained, having no boundary or edge, it would have neither beginning nor end: it would simply be. What place, then, for a creator?"

There's always a "but": "[A mathematical model] cannot answer the questions of why there should be a universe for the model to describe. Why does the universe go to all the bother of existing?" *Who says "why" means anything, except to us? Why can't the universe "simply* be"?

"In 1981 . . . I attended a conference on cosmology organized by the Jesuits in the Vatican. . . . The participants were granted an audience with the pope [John Paul II]. He told us that it was all right to study the evolution of the universe after the Big Bang, but we should not inquire into the Big Bang itself because that was the moment of Creation and therefore the work of God. I was glad then that he did not know the subject of the talk I had just given. . . . I had no desire to share the fate of GALILEO, with whom I feel a strong sense of identity, partly because of the coincidence of having been born exactly 300 years after his death!"

Judith Hayes, *senior writer for the* American Rationalist; *keeper of the Web site* The Happy Heretic (happyheretic.com); *author of a book by that name and another called* In God We Trust: But Which One? *Raised as a Catholic; began to question her faith after realizing her Hindu friend could not enter the Christian heaven, even if accompanied by a member. Enjoys mocking our Holy Bible for what she calls contradictions and absurdities.*

"The biblical account of Noah's Ark and the Flood is perhaps the most implausible story for fundamentalists to defend. Where, for example, while loading his ark, did Noah find penguins and polar bears in Palestine?"

"I can think of nothing, outside of sporting events or musical concerts, that could bring thousands of freethinkers together, and, further, bring them to their feet in wild applause and tears of joy. Can you? Even if you got the best freethinking orator around, someone like STEVE ALLEN, there is simply no way he could arouse our passions to that extent. 'Separation now!' just doesn't pack the same wallop as 'Soldiers for Jesus!'"

Hugh Hefner (1926–), *American playboy. Claims to be a direct descendant of William Bradford, a Mayflower Puritan. A species of rabbit (*Sylvilagus palustris hefneri*) is named in his honor.*

"Religion is a myth we have invented to explain the inexplicable. [The universe] is so beyond comprehension. What does it all mean—if it has any meaning at all? . . . I think anyone who suggests that they have the answer is motivated by the need to invent answers, because we have no such answers."— *Playboy, January. 2000 (Centerfold: Carol and Darlene Bernaola.)*

G. W. F. Hegel (1770–1831), *German philosopher. No one can understand a word he wrote. "Pure nonsense . . . a monument to German stupidity."—ARTHUR SCHOPENHAUER. "Reminiscent of the megalomaniac language of schizophrenics."—CARL JUNG.* "The sentiment underlying religion in the modern age [is] the sentiment God is dead."

So it didn't quite originate with NIETZSCHE: "God is, as it were, the sewer into which all contradictions flow."

Heinrich Heine (1797–1856), *German-Jewish poet. Converted to Protestantism in order to teach university, one of many professions then closed to Jews. His leadership in the Young Germany movement led to his books being banned in Germany. In 1933 they were burned.*

"In dark ages people are best guided by religion, as in a pitch-black night a blind man is the best guide. . . . When daylight comes, however, it is foolish to use blind, old men as guides."

"If your right eye offends you, pluck it out / If your right arm offends you, cut it off / And if your reason offends you, become a Catholic."

"Where they burn books, they will, in the end, burn human beings too."—*The remark is engraved on the plaza in Berlin where the Nazis held public book burnings. Heine was in fact referring to the burning of the Koran by the Spanish Inquisition.*

Robert A. Heinlein (1907–1988), *American science-fiction author. A pioneer of "social" science fiction. Was active in* UPTON SIN-CLAIR'*s socialist End Poverty In California (EPIC) movement in the early 1930s. Later swung to the right. Wrote apologias for the McCarthy hearings. Mounted a petition drive to urge President Eisenhower not to stop nuclear weapons testing. (And built a bomb shelter under his house.) Worked on the 1964* BARRY GOLDWATER *campaign. But supported "free love." Was a nudist. A Goldwater nudist. Satirized religion, especially evangelical Christianity, in novels such as* I Will Fear No Evil *and* Job: A Comedy of Justice.

"Men rarely (if ever) managed to dream up a god superior to themselves. Most gods have the manners and morals of a spoiled child."

"A long and wicked life followed by five minutes of perfect grace gets you into Heaven. An equally long life of decent living and good works followed by one outburst of taking the name of the Lord in vain—then have a heart attack at that moment and be damned for eternity. Is that the system?"

Claude Adrien Helvetius (1715–1771), *French philosopher and poet. His main work,* De l'esprit, *argued that humans are essentially animals, our behavior and ethics spring from self-interest, free will is an illusion, and right and wrong are culturally relative. The book was condemned by the Church and publicly burned by the hangman. Which of course lit a fire under sales and led to multiple foreign translations. (Please, censor me, somebody! Would a little fatwah be too much to ask?) On the Eucharist wafer:*

"A man who believes that he eats his God we do not call mad; a man who says he is Jesus Christ, we call mad." *(Except for one. Are we mad?)*

Ernest Hemingway (1899–1961), *American author. Raised in a strict Congregationalist household, but "ceased to practise religion at the earliest possible moment" and "regarded organized religion as a menace to human happiness."*[8] *Converted to Catholicism when he married a devout Catholic, Pauline Pfeiffer, in 1927. Turned against Catholicism while reporting (as a left-wing pro-Republican) on the Spanish Civil War. This apostasy led directly to a series of misfortunes, accidents, and health problems, aggravated by his alcoholism. Lost his home in Key West, Florida—a gift from Pauline's parents—when he divorced her in 1940. Lost his estate in Cuba as a result of the Castro revolution. Committed suicide, by shotgun, shortly after that.*

"All thinking men are atheists."

"Religion is like an ice cold whiskey on a hot day." *As a thinking man, that sounds pretty good to me.*

Bobby Henderson, *founder in 2005 of the religion of Flying Spaghetti Monsterism. "Pastafarians" believe the universe was created by a divine, all-knowing mass of spaghetti and meatballs. All evidence pointing toward evolution was for some reason intentionally planted on Earth by the FSM. Prayers are concluded with the word "RAmen." Though inspired by divine revelation, FSM-ism arose as a response to the 2005 decision by the Kansas State Board of Education to require the teaching of intelligent design. Perhaps the first religion—certainly the first pasta-based religion—to exist entirely on the Internet. From his Open Letter to the Kansas School Board:*

"I am concerned . . . that students will only hear one theory of Intelligent Design. . . . If the Intelligent Design theory is not based on faith, but is another scientific theory, as is claimed, then you must also allow our theory to be taught, as it is also based on science, not on faith. . . . Furthermore, it is disrespectful to teach our beliefs without wearing His [FSM's] chosen outfit . . . full pirate regalia. . . . I think we can all look forward to the time when these three theories are given equal time in our science classrooms."

Heraclitus of Ephesus (ca. 535–475 B.C.E.), *Greek philosopher. Known as "The Obscure" and the "weeping philosopher" (vs. the "laughing"—or at least chortling—DEMOCRITUS).*

"The universe has been made neither by gods nor by men, but it has been, and is, and will be eternally." (*Also see* EMPEDOCLES *and* STEPHEN HAWKING.)

"A blow to the head will confuse a man's thinking; a blow to the foot has no such effect. This cannot be the result of [man having] an immaterial soul."

"Religion is a disease, but it is a noble disease."

George Herbert (1593–1633), *English metaphysical poet and priest. His older brother Edward, an important poet and philosopher, was often referred to as "the father of English deism." Making George its uncle.*

"The devil divides the world between atheism and superstition."

B. R. Hergenhahn (1934–), *professor emeritus of psychology, Hamline University, Minnesota.*

"After Newton, it was but a short step to removing God altogether."

Hendrik Hertzberg (1943–), *senior editor and principal political commentator,* New Yorker *magazine. Speechwriter for President Jimmy Carter.*

"Where is it written that if you don't like religion you are somehow disqualified from being a legitimate American? What was MARK TWAIN, a Russian? . . . If it is American to believe that God ordered Tribe X to abjure pork, or that he caused Leader Y to be born to a virgin, why is it suddenly un-American to doubt the prime mover of this unimaginably vast universe of quintillions of solar systems would likely be obsessed with questions involving the dietary and biosexual behavior of a few thousand bipeds inhabiting a small part of a speck of dust orbiting a third-rate star in an obscure spiral arm of one of millions of more or less identical galaxies?"

Alexander Herzen (1812–1870), *Russian writer. The "father of Russian socialism." Arrested at age 22 and sentenced to five years of exile for taking part in the singing of verses uncomplimentary to the czar. Was it worth it, Alex?*

"All religions have based morality on obedience, that is to say, on voluntary slavery. That is why they have always been more pernicious than any political organization. For the latter makes use of violence, the former—of the corruption of the will."

Moses Hess (1812–1875), *German-Jewish philosopher. Early apostle of socialism, collaborator of MARX and ENGELS (whom he converted to Communism). Probable author of Marx's "opium of the people" gag.*

"The first words through which God made himself known to man was that curse that the Bible loyally handed down . . . 'in the sweat of thy face shalt thou eat bread.' . . . The work 'by the sweat of thy face' has reduced man to slavery and misery; the 'activity out of love' [SPINOZA'S *phrase*] will make him free and happy."

John Heywood (1497–1580), *English writer who fled his country to avoid persecution of Catholics. Author of the proverbs "One good turn asketh another," "Haste maketh waste," "When the sun shineth, make hay," "Look ere ye leap," "Two heads are better than one," "Beggars should be no choosers," "All is well that ends well," "I know on which side my bread is buttered," "A penny for your thought," "Rome was not built in one day," "Better late than never," and "The more the merrier."*

So it didn't quite originate with NIETZSCHE: "The neer to the church, the further from God."

Joe Hill (Joel Emmanuel Hägglund, 1879–1915), *Swedish-born American migrant worker and legendary labor activist with the Industrial Workers of the World, a.k.a. the Wobblies. Executed for the murder of an ex-cop after a controversial trial that drew international attention. Some of his ashes, preserved by the IWW, were eaten by folk singer BILLY BRAGG at the suggestion of ABBIE HOFFMAN. From his song "The Preacher and the Slave," a parody of a Salvation Army hymn, "In the Sweet Bye and Bye":*

> "Long-haired preachers come out every night, / Try to tell you what's wrong and what's right; / But when asked how 'bout something to eat / They will answer with voices so sweet: / You will eat, bye and bye, / In that glorious land above the sky; / Work and pray, live on hay, / You'll get pie in the sky* when you die."

Hippocrates (c. 460–377 B.C.E.), *Greek physician. His Hippocratic Oath introduced the principles of patient confidentiality, record-keeping, and the injunction, ~~"First, make sure they can pay"~~ "First, do no harm."*

> "Men think epilepsy divine, merely because they do not understand it. We will one day understand what causes it, and then cease to call it divine. And so it is with everything in the universe."

> "Where prayer, amulets and incantations work it is only a manifestation of the patient's belief."

Brad Hirschfeld, *Orthodox rabbi. Vice president of the National Jewish Center for Learning and Leadership in New York City.*

> "Religion drove those planes into those buildings. It's amazing how good religion is at mobilizing people to do awful, murderous things. There is this dark side to it, and anyone who loves religious experience, including me, better begin to own that there is a serious shadow side to this thing."

---

* Origin of the phrase "pie in the sky."

Alfred Hitchcock (1899–1980), *British-American film director. Churchgoing Catholic; increasingly irreligious as he aged, like a fine British wine. Surveilled by the CIA for his use of uranium as a plot device in* Notorious *(starring Ingrid Bergman as Alicia Huberman). Once pointed out of a car window at a priest talking to a little boy, and said:*

> "That is the most frightening sight I have ever seen. [*Shouting out the window:*] Run, little boy. Run for your life!"

Christopher Hitchens (1949–), *British-born American journalist. Ex-socialist; self-described "liberal hawk" and "contrarian." The attitude of his former colleagues on the left—the result mainly of Hitchens's support for the 2003 invasion of Iraq—was perhaps best expressed by* ALEXANDER COCKBURN *in 2005: "What a truly disgusting sack of shit Hitchens is." Regular contributor to* Vanity Fair *and* Slate. *His books include the merciless* Missionary Position: MOTHER THERESA *in Theory and Practice (1995) and* God Is Not Great: The Case Against Religion *(2007). Has described his daily alcohol intake as enough "to kill or stun the average mule."*

> "I am not even an atheist so much as an antitheist; I not only maintain that all religions are versions of the same untruth, but I hold that the influence of churches, and the effect of religious belief, is positively harmful. Reviewing the false claims of religion I do not wish, as some sentimental materialists affect to wish, that they were true. I do not envy believers their faith. I am relieved to think that the whole story is a sinister fairy tale; life would be miserable if what the faithful affirmed was actually true. . . . There may be people who wish to live their lives under a cradle-to-grave divine supervision; a permanent surveillance and monitoring. But I cannot imagine anything more horrible or grotesque."

> "Just consider for a moment what their [the devout's] heaven looks like. Endless praise and adoration, limitless abnegation and abjection of self; a celestial North Korea."

> "A true believer . . . must also claim to have at least an inkling of what that Supreme Being desires. I have been called arrogant

myself in my time . . . but to claim that I am privy to the secrets of the universe and its creator—that's beyond my conceit."

"A JERRY FALWELL clone named Bailey Smith observed that 'God Almighty does not hear the prayers of a Jew.' This is the only instance known to me of an anti-Semitic remark having a basis in fact."

"I, too, have strong convictions and beliefs and value the Enlightenment above any priesthood or any sacred fetish-object. It is revolting to me to breathe the same air as wafts from the exhalations of the madrasahs . . . or the sermons of Billy Graham and Joseph Ratzinger. But these same principles of mine also prevent me from wreaking random violence on the nearest church, or kidnapping a Muslim at random and holding him hostage . . . or making a moronic spectacle of myself threatening blood and fire to faraway individuals who may have hurt my feelings. The babyish rumor-fueled tantrums that erupt all the time, especially in the Islamic world, show yet again that faith belongs to the spoiled and selfish childhood of our species."

"Time spent in arguing with the faithful is, oddly enough, almost never wasted. The argument is the origin of all arguments; one must always be striving to deepen and refine it; MARX was right he stated in 1844 that 'the criticism of religion is the premise of all criticism.'"

Adolf Hitler (1889–1945). *The Nazis took pains to portray Jesus as a blond Aryan and certainly no Jew. Hostile to Christianity?*

"I am now as before a Catholic and will always remain so."

"I believe that I am acting in accordance with the will of the Almighty Creator; by defending myself against the Jews, I am fighting for the Lord. . . . I would like to thank Providence and the Almighty for choosing me of all people. . . ."

"Loyalty and responsibility toward the people and the Fatherland are most deeply anchored in the Christian faith."

"I have followed the Church in giving our party program the character of unalterable finality, like the Creed. . . . The Church has realized that anything and everything can be built up on a document of that sort, no matter how contradictory or irreconcilable with it. The faithful will swallow it whole, so long as logical reasoning is never allowed to be brought to bear on it."

Dr. Franklin Littell, chairman of the religion department, Temple University: "The Holocaust was, of course, the bitter fruit of long centuries of Christian teaching about the Jewish people."

Peter de Rosa, former Jesuit priest and theologian: "In 1936 . . . Hitler assured his lordship [Bishop Berning of Osnabruch] there was no fundamental difference between National Socialism and the Catholic Church. Had not the church, he argued, looked on Jews as parasites and shut them in ghettos? 'I am only doing,' he boasted, 'What the church has done for fifteen hundred years, only more effectively.'"

Thomas Hobbes (1588–1679), *seminal English political philosopher. Fled to France after angering both the Puritans and the monarchists with his idea of the social contract, which denied the divine right of kings. Fled back to England after attacking the papacy. (A war of one against all, it would seem.) Accused of atheism by the clergy—some of whom wanted him burned as a heretic—for writing that all substances are material and therefore God must be material. (What material should have been the question. My guess would be stainless steel or titanium.) Parliament began but eventually dropped an atheism action against him.*

"Religions are like pills, which must be swallowed whole without chewing."

"Immortality is a belief grounded upon other men's sayings, that they knew it supernaturally; or that they knew those who knew them that knew others that knew it supernaturally."

Eric Hoffer (1902–1983), *American social thinker and, until age 65, longshoreman: dubbed "the Longshoreman Philosopher." No university education. The best-known of his ten books,* The True Believer *(1951), a study of political and religious mass movements, argued that self-righteousness and fanaticism are rooted in self-hatred, self-doubt, and insecurity; noted how*

readily, for example, fanatical Nazis became fanatical Communists and fanatical Communists became fanatical anti-Communists. "For the true believer the substance of the mass movement isn't so important as that he or she is part of that movement."[1]

"[A] substitute is usually embraced with vehemence and extremism, for we have to convince ourselves that what we took as second choice is the best there ever was. . . . Faith is to a considerable extent a substitute for the lost faith in ourselves . . . for the self-confidence born of experience and skill. Where there is the necessary skill to move mountains there is no need for the faith that moves mountains."

"The less justified a person is in claiming excellence for their own self, the more ready they are to claim all excellence for their nation, their religion, their race or their holy cause."

"Self-righteousness is a loud din raised to drown the voice of guilt within us."

"The savior who wants to turn men into angels is as much a hater of human nature as the totalitarian despot who wants to turn them into puppets."

"The devout are always urged to seek the absolute truth with their hearts and not their minds."

"When we debunk a fanatical faith or prejudice, we do not strike at the root of fanaticism. We merely prevent its leaking out at a certain point, with the likely result that it will leak out at some other point."

"The opposite of the religious fanatic is not the fanatical atheist but the gentle cynic who cares not whether there is a god or not."

"Our passionate preoccupation with the sky, the stars, and a God somewhere in outer space is a homing impulse. We are drawn back to where we came from."

"The devil personifies not the nature that is around us but the nature that is within us—the infinitely ferocious and cunning prehuman creature that is still within us, sealed in the subconscious cellars of the psyche."

Abbie (Abbott Howard) Hoffman (1936–1989), *American social and political activist. Cofounder in 1967 of the Youth International Party ("Yippies"). At his sentencing as a member of the Chicago Seven for conspiracy and inciting to riot at the 1968 Democratic National Convention, he suggested the judge try LSD and offered to set him up with a dealer he knew.*

"Sacred cows make the tastiest hamburger." . . . *Medium-well, please.*

Nicholas von Hoffman, *American journalist-columnist* (New York Observer, Washington Post). *His books include* Hoax: Why Americans Are Suckered by White House Lies *(2004). Descendant of a sixteenth-century Anabaptist prophet, Melchior Hoffman.*

"Islam and Christianity both have a sex fixation: Practitioners can't get enough of it, even as they despise the thought of it. Their self-inflicted contradictions drive them crazy, and so they drive us non-believers nuts trying to take away our dirty pictures and our evil Web sites. . . . And the stuff they believe makes your ordinary witchdoctor look like Louis Pasteur or Jonas Salk. Submicroscopic homunculi running around inside a single cell, dinosaurs in the Garden of Eden. . . ."

Baron d'Holbach (Paul Henri Thiry, 1723–1789), *French philosopher and encyclopaederiste. Kept a twice-weekly salon in Paris whose regulars included* HELVÉTIUS, DIDEROT, *and other encyclos. Guys like Adam Smith,* DAVID HUME, *and* EDWARD GIBBON *often dropped by, probably just for the free food and booze. Author of* Christianity Unveiled *and* The System of Nature, *in which he denied (are you sitting down?) the very existence of God. The Catholic Church in France threatened to withdraw financial support from the crown unless it suppressed the book, which was too much even for* VOLTAIRE *and* FREDERICK THE GREAT— *and they weren't easily shocked.*

"Theology is but the ignorance of natural causes reduced to a system."

"As man exists in nature, I am not authorized to say that his formation is above the power of nature."

R. J. Hollingdale (1930–2001), *English scholar and translator of* NIETZSCHE, SCHOPENHAUER, *and other krauts.*

"[For demonstrating that God is dead] I admit that the generation which produced Stalin, Auschwitz and Hiroshima will take some beating; but the radical and universal consciousness of the death of God is still ahead of us; perhaps we shall have to colonize the stars before it is finally borne in upon us that God is not out there."

Oliver Wendell Holmes Sr. (1809–1894), *American writer, poet, and physician.*

"Men are idolaters, and want something to look at and kiss, or throw themselves down before; they always did, they always will; and if you don't make it of wood, you must make it of words."

"Rough work, iconoclasm, but the only way to get at truth."

"The man who is always worrying about whether or not his soul would be damned generally has a soul that isn't worth a damn."

"The mind, once expanded to the dimensions of larger ideas, never returns to its original size."

Oliver Wendell Holmes Jr. (1841–1935), *U.S. Supreme Court justice.*

"On the whole I am on the side of the unregenerate who affirm the worth of life as an end in itself, as against the saints who deny it."

Sidney Hook (1902–1989), *American philosopher. Headed the philosophy department at NYU for decades. A Marxist and cheerleader for the Soviet Union early in his career; later a passionate anti-Communist who*

helped create two secretly CIA-funded anti-Soviet organizations during the Cold War. Insisted he remained a socialist.

"Religious tolerance has developed more as the consequence of the impotence of religions to impose their dogmas on each other than as a consequence of spiritual humilty." *Or as W. H. Auden said:* "The only reason the Protestants and Catholics have given up the idea of universal domination [*not all of them!*] is because they've realised they can't get away with it."

Rev. Webster "Kit" Howell (1952–1996), *American Unitarian Universalist minister.*

"When we try to put God into the position of being a puppeteer who either pulls strings to make events happen or chooses to sit back, the suffering and evil in this world do become God's responsibility, and we can rightly accuse God of being a dysfunctional parent. But then, this is a child's view of God. . . . Why don't we grow up, access the divine within us (our reason, our compassion, our love), and do something about it?"

"What if God is like your mind—or your whole nervous system—in relation to your body . . . busy organizing matter into flesh. . . . Do we blame it for our ingrown toenails—and say, 'Mind is a jerk!'? Do we blame our mind when we develop ovarian cysts? . . . Why do things fall apart is the wrong question. The real wonder is: Why do things hang together? It is a divine miracle."

Huang Po (d. 850 c.e.), *influential Chinese master of Chan Buddhism. His message: Stop conceptualizing! Just say no.*

"The foolish reject what they see and not what they think; the wise reject what they think and not what they see."

Elbert Hubbard (1856–1915), *American philosopher. Edited and published a satirical magazine,* The Philistine, *which was bound in brown butcher paper because, said Hubbard, it had meat inside. Founded Roycroft,*

*an arts and crafts community in upstate New York, which became a meeting
site for radicals, freethinkers, suffragists and, one supposes, subversive weavers
and potters. Hubbard and his wife were killed in the sinking of the Lusitania
by a German U-boat. L. RON HUBBARD was a nephew.*

> "A mystic is a person who is puzzled before the obvious, but
> who understands the nonexistent."

> "Heaven: The Coney Island of the Christian imagination."

> "Who are those who will eventually be damned? Oh, the others,
> the others, the others!" *(Maybe that's what SARTRE meant. . . .)*

L. Ron (Lafayette Ronald) Hubbard (1911–1986,

if you believe he died), *American science-fiction writer and founder of
Scientology. Claimed to be a nuclear physicist on the basis of a one-year college course for which he received an F. During his stint in the Navy, he was
rated "unsatisfactory for any assignment" and "not temperamentally fitted for
independent command." Then there's Xenu, "the alien ruler of the 'Galactic
Confederacy' who, 75 million years ago, brought billions of people to Earth in
DC-8-like spacecraft, stacked them around volcanoes and blew them up with
hydrogen bombs. Their souls then clustered together and stuck to the bodies of
the living, and continue to wreak chaos and havoc today."[1] But is every word
of this Hubbard teaching* literally *true or are parts of it metaphorical?*

> "Writing for a penny a word is ridiculous. If a man really wants
> to make a million dollars, the best way would be to start his own
> religion." *(Reportedly uttered words to this effect on five separate
> occasions.)*

L. Jack Huberman (1954–), *Canadian-Jewish-American
writer. Professed "militant atheist" and "militant narcissist." Attended synagogue every Saturday for two weeks, around age 11, because some friends
were doing it. Soon resumed arguing the rationalist position with his Torah
teachers and Hasidic sidewalk missionaries. From his blog:*

> "The main debate seems to be whether the Rapture occurs at
> the beginning of the Tribulation or the end. Personally, I believe

the Rapture comes after the Seduction and during the Consummation, which of course is followed by the Resurrection and Repenetration."

Langston Hughes (1902–1967), *African-American writer and poet. His 1932 poem "Goodbye Christ," written during a visit to the Soviet Union, caused a scandal; years later, lectures by the "atheistic communist" and "notorious blasphemous poet" were still being disrupted by protesting evangelicals, right wingers, and assholes.*

"Goodbye, Christ Jesus Lord God Jehova / Beat it on away from here now / Make way for a new guy with no religion at all / A real guy named / Marx Communist Lenin Peasant Stalin Worker ME— / I said, ME!"

Robert Hughes (1938–), *Australian art critic.* Time *magazine art critic from 1970 on. Elected as one of 40 "living national treasures" by the Australian media in 1997.*

On attending a Catholic school:"Every time you wanked, it was a slaughter of future Catholics so small that a hundred of them could dance, or at least wiggle, on the head of a pin. . . . The notion that some small part of the cosmic order hung on our teenage willies was a heavy load for us young soldiers in St. Ignatius' army of Christ. In some of us, including Private Hughes, it induced the kind of suffocating guilt that led to scepticism: if God was so busy counting sperm, and so apparently unconcerned with preventing the world's famines, epidemics and slaughters, was He worth worshipping? Was He there at all? No answer from the altar."

"Nobody—except those who believe, on no evidence at all, that an immortal soul really is implanted in the embryo at the moment of conception, thus endowing it with complete humanity—can say at what point an embryo turns into a human being. The innocence of fetuses is not in doubt. But it is irrelevant: lettuces are innocent too."

Victor Hugo (1802–1885), *French novelist. Professed "freethinker."*
*Violently anticlerical. (The feeling was mutual: Hugo counted 740 attacks
on* Les Misérables *in the Catholic press.) His will stipulated that he be
buried without crucifix or priest. (It does get crowded in there with a priest.
Still: When you go, take one with you.)*

> "Religion is nothing but the shadow cast by the universe on human intelligence."

> "There is in every village a torch: the schoolmaster—and an extinguisher: the parson."

> "When you tell me that your Deity made you in his own image, I reply that he must have been very ugly."

> "God made himself man: granted. The Devil made himself woman." *(We don't whitewash here at* The Quotable Atheist.*)*

David Hume (1711–1776), *incredibly important Scottish philosopher
and historian. His essay* Of Superstition and Religion *underlies most
subsequent secular thought about religion—which Hume described as
"nothing but sick men's dreams." His criticism of the "design argument" for
the existence of God is widely accepted (not widely enough, of course) as
having killed the argument for good. Was tried for and acquitted of heresy;
the defense was that, as an atheist, he lay outside the jurisdiction of the
church. Was nonetheless denied teaching positions in Scotland, and his later
works on religion were held from publication until after his death.*

> "The Christian religion not only was at first attended with miracles, but even at this day cannot be believed by any reasonable person without one."

> "Men dare not avow, even to their own hearts, the doubts which they entertain on such subjects. They make a merit of implicit faith; and disguise to themselves their real infidelity."

> "If there is a designer he must take credit for the flaws in his creation. Flaws in the creation directly reflect flaws in the creator. If there is a flaw in the creator then he cannot be all powerful."

Dave Hunt (1926–), *American evangelical Protestant minister. Author of some three dozen books attacking Catholicism, Islam, Mormonism, Calvinism, New Age spirituality, and Protestant-Catholic ecumenism. In A* Woman Rides the Beast, *he identified the Catholic Church as the "Whore of Babylon" prophesied in Revelation. Let us bend his words to our purpose.*

> "Almost their [the Crusaders'] first act upon taking Jerusalem 'for Holy Mother Church' was to herd all of the Jews into the synagogue and set it ablaze. . . . The Secretary to the Inquisition in Madrid from 1790–1792 . . . estimated that in Spain alone the number of condemned exceeded 3 million, with about 300,000 burned at the stake. . . . Nor have the descendants of Aztecs, Incas, and Mayas forgotten that Roman Catholic priests, backed by the secular sword, gave their ancestors the choice of conversion (which often meant slavery) or death."

Aldous Huxley (1894–1963), *English novelist (*Brave New World, Eyeless in Gaza, Crome Yellow*), poet, and essayist. Moved to Hollywood, California—spiritual capital of the modern world. Soon hanging with J.* KRISHNAMURTI *and other gurus, swamis, and homies. . . . Posthumously a guru himself to hippie freaks through his book* The Doors of Perception, *which described his experiences with mescaline and LSD (and inspired the name of Jim Morrison's band). Grandson of* THOMAS HUXLEY, *brother of* JULIAN.

> "You never see animals going through the absurd and often horrible fooleries of magic and religion. . . . Dogs do not ritually urinate in the hope of persuading heaven to do the same and send down rain. Asses do not bray a liturgy to cloudless skies. Nor do cats attempt, by abstinence from cat's meat, to wheedle the feline spirits into benevolence. Only man behaves with such gratuitous folly. It is the price he has to pay for being intelligent but not, as yet, quite intelligent enough."

> "Maybe this world is just another planet's hell."

Sir Julian Huxley (1887–1975), *English biologist. Grandson of* THOMAS HUXLEY.

"Newton showed that gods did not control the movements of the planets; LAPLACE in a famous aphorism affirmed [*to Napoleon*] that astronomy had no need of the god hypothesis; DARWIN and Pasteur between them did the same for biology. . . . Operationally, God is beginning to resemble not a ruler but the last fading smile of a cosmic Cheshire cat."

Thomas Henry Huxley (1825–1895), *English biologist. Demonstrated that human and ape anatomy were fundamentally similar in every detail. Dubbed "Darwin's bulldog" for defending his work against the ferocious attacks of the dog-collared set (the clergy). Coined the word "agnostic." Grandfather of* ALDOUS *and* JULIAN HUXLEY, *who seem to have inherited his acquired characteristics, atheistically speaking.*

"[I am] inclined to think that not far from the invention of fire must rank the invention of doubt."

"Skepticism is the highest duty and blind faith the one unpardonable sin."

"The known is finite, the unknown is infinite; intellectually we stand on an islet in the midst of an illimitable ocean of inexplicability. Our business in every generation is to reclaim a little more land."

"Extinguished theologians lie about the cradle of every science as the strangled snakes beside that of Hercules."

After reading Darwin's Origin of Species: "How stupid of me not to have thought of that!" *(Compared to still not accepting it today? Let's make that 10 on a stupid-scale of 10 . . .)*

SECULAR * INFIDEL * NONBELIEVER * HUMANIST * RATIONALIST * FREETHINKER * AGNOSTIC * GODLESS * HERETIC * ATHEIST

Michael Ignatieff (1947–), *Canadian journalist/scholar/politician. Elected in 2006 as a Liberal member of Parliament; candidate for party leadership.*

> "Secular faiths such as Marxism left much less room for doubt than religion. Religion knows all about doubt. Since God's ways are unknowable, religion can endure only if it finds a place for self-questioning. Prayer itself is a questioning dialogue with God."

Robert G. Ingersoll (1833–1899), *American politician and atheist, uh, crusader; the most famous atheist of his day. (In the 1880s the three highest paid speakers in the U.S. were atheists—Ingersoll, THOMAS HUXLEY, and MARK TWAIN.) Known as "the Pagan prophet." Son of a minister. ("I'll show him . . ."?) "When arrested under the Comstock laws [see MARGARET SANGER] for mailing birth control information, Ingersoll argued in his defense that the Bible is full of sexually explicit passages, so it, too, should be barred from the mail system."*[2]

> "Every fact is an enemy of the church. Every fact is a heretic. Every demonstration is an infidel. Everything that ever really happened testifies against the supernatural."

"The clergy know that I know that they know that they do not know."

"If a man would follow, today, the teachings of the Old Testament, he would be a criminal. If he would follow strictly the teachings of the New, he would be insane."

"In nature there are neither rewards nor punishments—there are consequences."

"Our civilization is not Christian. It does not come from the skies. It is not a result of 'inspiration.' It is the child of invention, of discovery, of applied knowledge—that is to say, of science."

"When worship shall consist in doing useful things; when religion means the discharge of obligations to our fellow-men, then, and not until then, will the world be civilized."

"Hands that help are far better then lips that pray."

"Religion supports nobody. It has to be supported. . . . It is a perpetual mendicant. It lives on the labors of others, and then has the arrogance to pretend that it supports the giver."

"The hope of science is the perfection of the human race. The hope of theology is the salvation of a few, and the damnation of almost everybody."

"With soap, baptism is a good thing."

Mary Jean Irion, *radical American theologist, poet, and University of Connecticut professor. Author of* From the Ashes of Christianity: A Post–Christian View *(1968).*

"Christianity . . . has been over for a hundred years now. . . . When something even so small as a lightbulb goes out, the eyes for a moment still see it; and a sound after it is made will have, in the right places, an echo. So it is not at all strange that when something so huge as a world religion goes out, there remains for a century or more in certain places some notion that it is still there." *Indeed, it may flare up brightly just before the end.*

"Faith is not making religious-sounding noises in the daytime. It is asking your inmost self questions at night—and then getting up and going to work."

Hirobumi Ito (1841–1909), *modern Japan's 1st, 5th, 7th, and 10th prime minister. A samurai who studied at University College, London.*

"I regard religion itself as quite unnecessary for a nation's life; science is far above superstition; and what is religion, Buddhism or Christianity, but superstition, and therefore a possible source of weakness to a nation?" . . . *a nation whose destiny is to rule all of Asia? (Also see D. T. SUZUKI and BRIAN DAIZEN VICTORIA.)*

William James (1842–1910), *American psychologist/philosopher. Brother of novelist Henry James. Author of the classic* The Varieties of Religious Experience. *Yearned for a religious experience himself, but couldn't get hold of one, despite attending church faithfully. Actually found it impossible to pray. Well, some atheists have to be dragged in kicking and screaming, and, kept restrained, gagged, and sedated throughout the reindoctrination period.*

"The arguments for God's existence have stood for hundreds of years with the waves of unbelieving criticism breaking against them, never totally discrediting them in the ears of the faithful, but on the whole slowly and surely washing out the mortar from between their joints."

"The God whom science recognizes must be a God of universal laws exclusively, a God who does a wholesale, not a retail business."

"Medical materialism finishes up SAINT PAUL by calling his vision on the road to Damascus a discharging lesion of the occipital cortex, he being an epileptic. It snuffs out Saint Teresa as an hysteric, Saint Francis of Assisi as an hereditary degenerate. George Fox's . . . pining for spiritual veracity, it treats as a symptom of a disordered colon. CARLYLE's organtones of misery it accounts for by a gastroduodenal catarrh."

Thomas Jefferson (1743–1826), *third U.S. president. Deist (see* JOHN ADAMS) *and adamant church-state separationist. "No other statesman of his time could match Jefferson in his hatred of the established faith," wrote a biographer. "Went so far as to edit the gospels, removing the miracles and mysticism of Jesus, leaving only what he deemed the correct moral philosophy."*[9] *Engraved on the Jefferson Memorial is his vow of "eternal hostility against every form of tyranny over the mind of man"—without mention that he said it while denouncing those who wanted to establish Christianity as the official religion.*

"Question with boldness even the existence of a god."

"I have recently been examining all the known superstitions of the world, and do not find in our particular superstition (Christianity) one redeeming feature. They are all alike, founded upon fables and mythologies."

"The Christian god is a three headed monster; cruel, vengeful and capricious. . . . One only needs to look at the caliber of people who say they serve him. They are always of two classes: fools and hypocrites."

"The day will come when the mystical generation of Jesus . . . in the womb of a virgin will be classed with the fable of the generation of Minerva in the brain of Jupiter."

"I am a Materialist; he [Jesus] takes the side of Spiritualism; he preaches the efficacy of repentance toward forgiveness of sin; I require a counterpoise of good works to redeem it."

"Christianity is the most perverted system that ever shone on man. . . . perverted into an engine for enslaving mankind . . . a mere contrivance [for the clergy] to filch wealth and power to themselves."

"In every country and in every age the priest has been hostile to liberty, he is always in allegiance with the despot, abetting his abuses in return for protection of his own. . . . History I believe furnishes no example of a priest-ridden people maintaining a free civil government. . . . Political as well as religious leaders will always avail themselves [of public ignorance] for their own purpose."

"State churches that use government power to support themselves and force their views on persons of other faiths undermine all our civil rights. . . . Erecting the 'wall of separation between church and state,' therefore, is absolutely essential in a free society."

Penn Jillette (1955–), *American magician. Described by NPR as "the taller, louder half of the magic and comedy act Penn and TELLER," who also host a Showtime cable series,* Bullshit! *on which they debunk pseudo-science, psychics, etc. "Penn is such an ardent atheist he refuses to go to weddings," said one magazine profile. Said to wear his "Team Satan 666" T-shirt everywhere. Asked by Donny and Marie Osmond (devout Mormons) for autographs, Penn wrote "There is no god"; Teller wrote, "He's right."*

"I'm beyond atheism. Atheism is not believing in God. That's easy—you can't prove a negative, so there's no work to do. . . . I'm saying, 'I believe there is no God.' Having taken that step, it informs every moment of my life. I'm not greedy. I have love, blue skies, rainbows and Hallmark cards, and that has to be enough . . . it just seems rude to beg the invisible for more." *(Not even better Hallmark cards?)*

"Believing there's no God means I can't really be forgiven except by kindness and faulty memories. That's good . . . I have to try to treat people right the first time around."

"When people over seven years old have imaginary friends, there's going to be trouble . . . they're going to be killing somebody."

Billy Joel (1949–), *American singer/songwriter/pianoman. Six-time Grammy winner. His atheist status has come "under review" at celebatheist.com over allegations of saying "God bless you," among other un-atheist statements. Readers are asked to watch for and report any signs of apostasy.*

"I wasn't raised Catholic, but I used to go to Mass with my friends, and I viewed the whole business as a lot of very enthralling hocus pocus. There's a guy hanging upon the wall . . . nailed to a cross and dripping blood and everybody's blaming themselves for that man's torment, but I said to myself, 'Forget it. I had no hand in that evil. I have no original sin. . . . I pass on all of this.' . . . I had some Jewish guilt in me already . . . so I knew I definitely had no room for Catholic guilt too. . . . Then my mother took my sister and me to an Evangelical church. . . . I was baptised there at the age of twelve, and it was strictly hallelujah time. But one day the preacher is up in the pulpit unfolding a dollar bill saying, 'This is the flag of the Jews.' Whoa, fella! We left that flock. . . . I gradually decided that just because I didn't have or couldn't find the ultimate answer didn't mean I was going to buy the religious fairytale."

Ellen Johnson, *president of American Atheists.*

"American Atheists has always encouraged the public to read both the Old and New Testaments from cover to cover. Many people become atheists after reading the Bible."

Sonia Johnson (1936–) *American feminist activist. Excommunicated by the Mormon Church in the late 1970s for her pro-Equal Rights Amendment activities (which included founding Mormons for ERA, which sounds like one of those "world's thinnest books" jokes). Author of* From Housewife to Heretic *(1981). Presidential candidate of Barry Commoner's U.S. Citizens Party in 1984, losing to Ronald Reagan by a mere 54,383,319 votes.*

"One of my favorite fantasies is that next Sunday not one woman, in any country of the world, will go to church. If women simply stop giving our time and energy to the institutions that oppress, they cease to be."

Terry Jones (1942–), *British comedian, film director, children's book author, and popular historian; member of the Monty Python comedy team. Directed the Python movies* Life of Brian, The Meaning of Life, *and (with* TERRY GILLIAM) Monty Python and the Holy Grail. *Claims to have written* Starship Titanic—*a novel based on a computer game by* DOUGLAS ADAMS—*entirely while in the nude. From his 2004 op-ed piece "George, God here . . .":*

"I want you to stop this Iraq thing, George." "But you told me to do it, God!" "No I didn't, George . . ." "But you did!" . . ."Listen, you ignorant little pinch-eyed Billy Graham convert! Can't you get it into your head that I'm God and I'm telling you to stop all this 'pre-emptive strike' nonsense! Stop destroying Iraq! . . . You're leading the world into unbelievable chaos and horror!" . . . "You speak through me, God, not the other way round! Is that clear?" "Yes, Mr. President."

S. T. Joshi (1958–), *Indian-American literary scholar and author. Editor of* Atheism: A Reader *(2000), a collection of writings by prominent atheists. Author of* God's Defenders: What They Believe and Why They Are Wrong *(2003), which attacks and seriously injures C. S. Lewis, G. K. Chesterton, T. S. Eliot, and other men of faith and initials.*

"The atheist, agnostic, or secularist . . . should not be cowed by exaggerated sensitivity to people's religious beliefs. . . . Those who advocate a piece of folly like the theory of an 'intelligent creator' should be held accountable for their folly; they have no right to be offended for being called fools until they establish that they are not in fact fools."

James Joyce (1882–1941), *Irish novelist. The artist as a young man rejected all religion and described Catholicism as "black magic."*

"There is no heresy or no philosophy which is so abhorrent to the church as a human being."

"The artist, like the God of the creation, remains within or behind or beyond or above his handiwork, invisible, refined out of existence, indifferent, paring his fingernails." *(About time— His fingernails were* gross!*)*

Carl Jung (1875–1961), *Swiss psychiatrist. Christian enough, God knows—and yet:*

"Religion is a defense against the experience of God."

"It is only through the psyche that we can establish that God acts upon us, but we are unable to distinguish whether these actions emanate from God or from the unconscious. We cannot tell whether God and the unconscious are two different entities."

Franz Kafka (1883–1924), *Austrian-Czech-Jewish novelist. Patron saint of alienation and paranoia—in short, of the kafkaesque.*

"We are sinful not merely because we have eaten of the tree of knowledge, but also because we have not eaten of the tree of life."

Joachim Kahl (1941–), *German author of* The Misery of Christianity: A Plea for Humanity without God *(1968). Credited his training in theology for converting him into "an open and pugnacious atheist." All too familiar story.*

"What after all is the cross of Jesus Christ? It is nothing but the sum total of a sadomasochistic glorification of pain. Christianity has not failed the ideals of its founder: Christ has failed. Corrupt in its very essence, the gospel of Christ alone has persecuted Jews, defamed the female and suppressed sexuality."

Azam Kamguian (1958–), *Iranian secularist writer and women's rights activist. Arrested and imprisoned for political activities under both the Shah and the mullahs; the latter subjected her to constant torture. Fled to Britain. Founded the Committee to Defend Women's Rights in the Middle East. Excoriates liberal "cultural relativists" who shy from criticizing Islam.*

"[To say] we must respect people's culture and religion, however despicable [is] absurd. . . . If a culture allows women to be mutilated and killed to save the family's 'honour,' it cannot be excused."

Wendy Kaminer, *American law professor and feminist writer. Her books include* Sleeping with Extra-Terrestrials: The Rise of Irrationalism and Perils of Piety. *Her 1996 article "The Last Taboo" dared to suggest that what America needs is not more religiosity or spirituality but more atheism! Prepare the faggots.*

"Americans have become fascinated by angels and 'out of body' experiences and seem to be discarding the habit of critical thinking. . . . The dissemination of pseudoscience, including such things as the fascination with near-death experiences and the growing belief by Americans—34 percent of them—in reincarnation are dangerous. They help to break down the standards of reason."

"People who believe that God exists and heeds their prayers have probably waived the right to mock people who talk to trees or claim to channel the spirits of Native Americans."

"If I were to mock religious belief as childish . . . I'd be excoriated as an example of the cynical, liberal elite responsible for America's moral decline. . . . I'd receive hate mail. Atheists generate about as much sympathy as pedophiles."

Immanuel Kant (1724–1804), *German philosopher. "Kant so fully demolished the last remaining philosophical proofs of God that* [Moses] Mendelssohn *called him 'the all destroyer,' and the name stuck."[2] Yet he continued to believe that because God inhabited a real world "out there" that we can never know, we might as well choose to believe in him. (Also see—but do not take—*Pascal's Wager.*)*

"Supreme Being is a mere ideal, the objective reality of which can neither be proved nor disproved by pure reason."

"The death of dogma is the birth of morality."

"*Sapere aude,* have the courage to know: this is the motto of the Enlightenment."

Sir Bernard Katz (1911–2003), *German-Jewish-born British biophysicist; 1970 Nobel winner for his work on nerve biochemistry.*

"Organized religion: The world's largest pyramid scheme."

Stuart Kauffman (1939–), *American theoretical biologist at the University of Calgary; researcher into the origins of life. Founder of a company that applies the science of complexity to business management problems. Maintains that self-organization is as important as Darwinian selection in explaining biological complexity.*

"Who seeing the snowflake, who seeing simple lipid molecules cast adrift in water forming themselves into cell-like hollow lipid vesicles, who seeing the potential for the crystallization of life in swarms of reacting molecules . . . can fail to entertain a central thought: if ever we are to attain a final theory in biology, we will surely, surely have to understand the commingling of self-organization and [Darwinian] selection. We will have to see that we are the natural expressions of a deeper order. Ultimately, we will discover in our creation myth that we are expected after all."
"Expected" by Anyone in particular? Kaufmann thinks the intelligent design "question" is "legitimate. I just worry about the methodologies, and hidden references to a creator."

Walter Kaufmann (1921–1980), *German philosopher and poet. Raised as a Lutheran; decided at age 11 to become a Jew because he could not believe in the Trinity or that Jesus was God. Leading translator of* NIETZSCHE *and other existentialists. Author of* The Faith of a Heretic. *"The only theism as profound as Buddha's atheism, Kaufman averred, was the theism of people like Job."* [2]

"The only theism worthy of our respect believes in God not because of the way the world is made but in spite of that."

"Few Christians would be in doubt what to think of a father who tortured his children for forty-eight hours because they did not agree with him or did not obey him; and if he had a great many children and had given only a few of them a single chance while offering the vast majority no opportunity at all to know his will, most people would consider this the epitome of an inhuman lack of love and justice."

Kayhan, *Iranian newspaper, directly under the supervision of the Office of the Supreme Leader.*

From an editorial urging cooperation with the Vatican in opposing a 1994 United Nations plan to stabilize the world's population (because we can easily cram 40 billion more good Muslims and Christians onto the planet—and God will provide):

"It is high time that scholars of all godly religions united to confront the forces of immorality in the present day under various names such as secularism, human rights, freedom of speech."

Garrison Keillor (1942–), *American humorist and host of the long-running public radio show* A Prairie Home Companion. *Raised in a family belonging to the fundamentalist Plymouth Brethren. Strong, good-looking, and above average.*

"My ancestors were Puritans from England. They arrived here in 1648 in hopes of finding greater restrictions than were permissible under English law at that time."

Florynce "Flo" Kennedy (1916–2000), *African-American civil rights attorney and feminist activist. One of the first black women to graduate from Columbia Law School (1951). Helped found NOW. Sued the Roman Catholic Church for political interference with abortion rights by a tax-exempt organization. Led a mass urination by women protesting a lack of women's bathrooms at Harvard. Wore a cowboy hat and pink sunglasses at all times. People magazine called her "the biggest, loudest and, indisputably, the rudest mouth on the [feminist-radical-political] battleground. . . ."*

"It's interesting to speculate how it developed that in two of the most anti-feminist institutions, the church and the law court, the men are wearing the dresses."

Jomo Kenyatta (1892–1978), *first prime minister and president of independent Kenya. Grandson of the village medicine man. Went to school in a Scottish mission center and was converted to Christianity at age 22 with the name John Peter.*

> "When the missionaries arrived, the Africans had the Land and the Missionaries had the Bible. They taught us how to pray with our eyes closed. When we opened them, they had the land and we had the Bible."

Johannes Kepler (1571–1630), *German astronomer, astrologer (part of his university studies in theology), and early writer of science-fiction stories. Discovered the mathematical laws of planetary motion. His mother, a healer and herbalist, was imprisoned for 14 months on charges of witchcraft before his legal efforts secured her release and permitted her swift return home by broomstick.*

> "When miracles are admitted, every scientific explanation is out of the question."

Jack Kevorkian (1928–), *a.k.a. "Dr. Death": American pathologist and right-to-die activist. Claimed to have helped at least 130 people to die. In 1999, began serving a 10- to 25-year prison sentence for second-degree murder in an assisted suicide, a videotape of which he gave to 60 Minutes. Saturday Night Live once lampooned Kevorkian's decision to purchase a pistol (for self-defense), saying "Alright, now he's just getting lazy."*

> *Identifying his God:* "Bach. At least I didn't make mine up."

Omar Khayyam (ca. 1048–1131), *Persian poet, philosopher, and scholar. His famous poem the* Rubayyat *"scoffs at theologians, laments the unknowability of the hereafter, and hails wordly pleasure as the only tangible goal."[3] The mullahs have issued a fatwah sentencing him to death retroactively. Just kidding. But verses of the* Rubayyat *proclaiming that "My life-long* *practice is to praise the Vine" and "I have drunk and drink now and will drink still" have been "officially reinterpreted" in Iran: He did not actually drink wine, which may only have symbolized mystical Sufi poetry or spiritual ecstasy.* ≈

And how do they explain this *away?* "The revelations of the devout and learned / Who rose before us and as prophets burned / Are all but stories. . . .

Ayatollah Ruhollah Khomeini (1901–1989), *Iranian religious-political leader.*

"There is no room for play in Islam. . . . It is deadly serious about everything." *"Deadly" is the word.*

Søren Kierkegaard (1813–1855), *Danish Christian existentialist philosopher/theologian. "By far the most profound thinker of the 19th century," according to* WITTGENSTEIN. *Repeatedly said he was incapable of faith, but believed in it, longed for it. The "leap of faith," transcending rationality, was his stupid idea. For Kierkie, "to have faith that God exists, without ever having doubted God's existence . . . would not be a faith worth having."[1] (Nor would it be with the doubt.) Apparently driven half-mad by uncertainty about how to pronounce the ø in Søren.*

"Christendom has done away with Christianity, without being aware of it. . . . The ideals of the New Testament have gone out of life. . . . [This] has been perceived by many. They like to give it this turn: the human race has outgrown Christianity."

Stephen King (1947–) *American horror novelist. Got a $2,500 advance for his first novel,* Carrie, *whose heroine's insane mother is one of the truest portraits of a zealous evangelical Christian ever written—not least with regard to the sexual origins of her zealotry. (See* KRAFFT-EBBING *and* BARING-GOULD.) *Doesn't own a cell phone.*

"The beauty of religious mania is that it has the power to explain everything . . . nothing is left to chance . . . logic can be happily tossed out the window."

Margaret Knight (1903–1983), *British psychologist and the most famous British atheist/humanist of the second half of the twentieth century. In 1955, she caused an SOC (storm of controversy) with her series of talks on BBC radio based on her book* Morals without Religion, *in which she offered advice on how to teach that very thing to defenseless children.*

"The fundamental opposition is between dogma and the scientific outlook. On the one side, Christianity and Communism, the two great rival dogmatic systems; on the other Scientific Humanism."

"If [a child] is normally intelligent, he is almost bound to get the impression that there is something odd about religious statements. If he is taken to church, for example, he hears that death is the gateway to eternal life, and should be welcomed rather than shunned; yet outside he sees death regarded as the greatest of all evils, and everything possible is done to postpone it. . . . The child soon gets the idea that there are two kinds of truth—the ordinary kind, and another, rather confusing and slightly embarrassing kind, into which it is best not inquire too closely. . . . Now all this is bad intellectual training. . . ."

Nanrei Kobori (1918–1998), *Japanese Zen Buddhist monk; Abbot of the temple of Ryokoin, Kyoto.*

"God is an invention of Man. So the nature of God is only a shallow mystery. The deep mystery is the nature of man."

Arthur Koestler (1905–1983), *Hungarian-Jewish-British social and political philosopher and novelist. All over the map in many senses: Communist turned fervent anti-Stalinist (his most famous novel,* Darkness at Noon, *is about the Soviet purges of the 1930s; contributed to* The God That Failed *[1950], a collection of testimonies by ex-Commies); served in the British Army and the French Foreign Legion; member of a zeppelin expedition to the North Pole in 1931; by age 40, had lived in Hungary, Austria, Germany, Palestine, the USSR, Britain, and France. Participated in* TIMOTHY LEARY's *early experiments with psychedelic drugs (1960). Tried to connect quantum theory and* JUNG'S *concept of synchronicity in* The Roots of Coincidence. *The*

Police album Ghost in the Machine *was named after book by Koestler about non-Darwinian evolution. His will endowed the chair of parapsychology at the University of Edinburgh. Lifelong atheist.*

> "Faith is a wondrous thing; it is not only capable of moving mountains, but also of making you believe that a herring is a race horse."

> "God seems to have left the receiver off the hook, and time is running out."

> "Scientists are peeping toms at the keyhole of eternity."

Jeremy Konopka, *possibly a Canadian computer science professor at the University of Regina. Definitely the contributor of this quote to alt.atheism.*

> "In the brain of every religious person there is a god-shaped vacuum."

Karl Kraus (1874–1936), *Austrian journalist, essayist, and renowned satirist. Traded in his matzohs for communion wafers at age 25, but 12 years as a Catholic was enough. Founded his own newspaper in which he attacked a variety of sacred cows and sheep.*

> "When a culture feels that its end has come, it sends for a priest."

Baron Richard von Krafft-Ebing (1840–1902), *Austro-German psychiatrist; author of* Psychopathia Sexualis *(1886), the famous study of sexual perversity which today is found in every nightstand drawer in America. Coined the terms* sadism *and* masochism.

> "We find that the sexual instinct, when disappointed and unappeased, frequently seeks and finds a substitute in religion." *Not in my case. . . .*

J. Krishnamurti (1895–1986), *Indian-born philosopher/teacher. Discovered as a teenager by the Theosophical Society and raised within the organization (largely by ANNIE BESANT), whose leaders believed him to be a prophesied World Teacher. Disavowed this destiny in his late twenties (just when Australian Theosophists were expecting him to arrive in Sydney literally walking on water), and spent the rest of his life traveling the world as an independent spiritual speaker and educator, refusing to accept followers and . . . attracting a large following. Addressed the United Nations 'and was awarded the UN Peace Medal at age 90. Close friends with ALDOUS HUXLEY and JOSEPH CAMPBELL.*

"Truth is a pathless land. . . . A belief is purely an individual matter, and you cannot and must not organize it. If you do, it becomes dead, crystallized; it becomes a creed, a sect, a religion, to be imposed on others."

"For centuries we have been spoon-fed by our teachers, by our authorities, by our books, our saints. We say, 'Tell me all about it . . .' and we are satisfied with their descriptions. . . . We are secondhand people."

"Truth is something living, moving, which has no resting place, which is in no temple, mosque or church, which no religion, no teacher, no philosopher, nobody can lead you to. . . . There is only you—your relationship with others and with the world— there is nothing else. . . . Neither your gods, nor your science can save you, can bring you psychological certainty; and you have to accept that you can trust in absolutely nothing."

Nicholas D. Kristof (1959–), New York Times *columnist and former Hong Kong, Beijing, and Tokyo bureau chief.*

"Despite the lack of scientific or historical evidence, and despite the doubts of Biblical scholars, America is so pious that not only do 91 percent of Christians says they believe in the Virgin Birth, but so do an astonishing 47 percent of U.S. non-Christians."

Stanley Kubrick (1928–1999), *American movie director. Atheist?*
The mystifying mysticism of 2001: A Space Odyssey *and the occultism of*
The Shining *would suggest not. In connection to the latter, Kubrick said he*
hoped ghosts did exist as it meant some sort of afterlife. We quote, you
decide.

> "The whole idea of god is absurd. If anything, 2001 shows that
> what some people call 'god' is simply an acceptable term for
> their ignorance. . . . This film is a rejection of the notion that
> there is a god; isn't that obvious?"

> "I'd be very surprised if the universe wasn't full of an
> intelligence of an order that to us would seem God-like. . . .
> There are approximately 100 billion stars in our galaxy alone
> [and] approximately 100 billion galaxies in just the visible
> universe . . . so it seems likely that there are billions of planets
> in the universe . . . where intelligent life . . . is hundreds of
> thousands or millions of years in advance of us. When you
> think of the giant technological strides that man has made in
> a few millennia—less than a microsecond in the chronology of
> the universe—can you imagine the evolutionary development
> that much older life forms have taken? They may have
> progressed from biological species, which are fragile shells for
> the mind at best, into immortal machine entities—and then,
> over innumerable eons, they could emerge from the chrysalis
> of matter transformed into beings of pure energy and spirit.
> Their potentialities would be limitless and their intelligence
> ungraspable by humans."

Milan Kundera (1929–), *Czech-French writer. Joined the ruling*
Czechoslovak Communist Party in 1948. Expelled for "antiparty activities"
in 1950. Readmitted in 1956. Reexpelled in 1970. Unbearable.

> "Man desires a world where good and evil can be clearly
> distinguished, for he has an innate and irrepressible desire to
> judge before he understands."

Paul Kurtz (1925–) *American philosophy professor and one-man Age of Enlightenment. Founder of the Council for Secular Humanism, the Center for Inquiry, the Committee for the Scientific Investigation of Claims of the Paranormal (CSICOP), and Prometheus Books, which specializes in secularist works. Editor in chief of* Free Inquiry *magazine; former copresident of the International Humanist and Ethical Union; president of the International Academy of Humanism; author or editor of over 40 books. Has labored to promote secularism as a positive choice, not merely religion-bashing—but acknowledges that that's the fun part. Or should.*

> "Homo religiosus invents religious symbols, which he venerates and worships to save him from facing the finality of his death and dissolution. . . . In the last analysis it is the theist who can find no ultimate meaning in this life and who denigrates it. . . . The theist can only find meaning by leaving this life for a transcendental world beyond the grave."

SECULAR * INFIDEL * NONBELIEVER * HUMANIST * RATIONALIST * FREETHINKER * AGNOSTIC * GODLESS * HERETIC * ATHEIST

Cathy Ladman (1955–), *American comedian/actress (lots of sitcoms).*

"All religions are the same: religion is basically guilt, with different holidays."

Julien Offray de La Mettrie (1709–1751), *French physician and philosopher. The outcry provoked by his* Natural History of the Soul, *which attributed mental phenomena to organic changes in the brain and nervous system, forced him to flee to Holland, where the publication of his even bolder book,* Man the Machine, *forced him to flee to Berlin, where he was appointed court reader by* FREDERICK, *who wasn't called "THE GREAT" for nothing.*

"Atheism is the only means of ensuring the happiness of the world, which has been rendered impossible by the wars brought about by theologians." *(In our era, for "theologians," read reverends and rebbes, imams and mullahs.)*

"The soul is only the thinking part of the body, and with the body it passes away. When death comes, the farce is over, so let us take our pleasure while we can." *True to his word, when his death came, it was from eating a vast quantity of* pâte aux truffes *at a banquet.*

Anne Lamott (1954–), *American novelist and nonfiction writer.*

"You can safely assume that you've created God in your own image when it turns out that God hates all the same people you do."

Kenneth V. Lanning, *FBI expert on "satanic, occult and ritualistic crime"; author of the 1992 FBI Report "Satanic Ritual Abuse." Commenting on popular beliefs that children by the thousands were being murdered in human sacrifices, babies were being bred and eaten, and Satanists were taking over America's day care centers (all of which had ended by the mid-1980s at the latest):*

"Many children in the United States . . . are severely psychologically, physically, and sexually traumatized by angry, sadistic parents or other adults. . . . The vast majority of [these] were abused by Christians."

"Nothing is more simple than 'the devil made them do it.' . . . Especially for those raised to religiously believe so, Satanism offers an explanation as to why 'good' people do bad things [and may] help to 'explain' bizarre and compulsive sexual urges. . . . The fact is that far more crime and child abuse has been committed by zealots in the name of God, Jesus and Mohammed than has ever been committed in the name of Satan."

Pierre Laplace (1749–1827), *French physicist, mathematician, and astronomer. To Napoleon Bonaparte who, after reading Laplace's Celestial Mechanics—which explained the universe purely in terms of natural causes—observed that it contained no mention of God:*

"Sire, I have no need of that hypothesis." *How cool is that?*

Anton Szandor LaVey (1930–1997), *American founder and high priest of the Church of Satan; author of The Satanic Bible (found in most hotel nightstand drawers). Dropped out of high school to work in circuses*

and carnivals. His interest in the occult led to his founding of the Church of Satan in 1966, henceforth known as "the year One A.S. (Anno Satanas)." Shaved his head and grew a devilish moustache and goatee. (That's just Satanic Leadership 101.) Viewed Satan merely as a symbolic and literary figure; based the religion on the materialism and individualism of such spawn of hell as NIETZSCHE *and* AYN RAND. *Allies and followers included Jayne Mansfield, Sammy Davis Jr., and* MARILYN MANSON.

> "Many of you have already read my writings identifying TV as the new God. . . . In previous centuries, the Church was the great controller, dictating morality, stifling free expression and posing as conservator of all great art and music. Today we have television dictating fashions, thoughts, attitudes, objectives as once did the Church, using many of the same techniques but doing it so palatably that no one notices. Instead of 'sins' to keep people in line, we have fear of being judged unacceptable by our peers (by not wearing the right running shoes, not drinking the right kind of beer or wearing the wrong kind of deodorant), and fear of imposed insecurity concerning our own identities. Borrowing the Christian sole salvation concept, television tells people that only through exposure to TV can the sins of alienation and ostracism be absolved."

D. H. (David Herbert) Lawrence (1885–1930), *English writer. "At the time of his death, his public reputation was that of a pornographer who had wasted his considerable talents."*[1] *(Lady Chatterley's Lover [1928] was only published in 1960, and even then the publisher was tried for obscentity.) This, however, was an extremely religious, indeed Christian, pornographer, in his carnal, Laurentian way.*

> "I worship Christ . . . Jehovah . . . Pan . . . Aphrodite. But I do not worship hands nailed and running with blood."

> "Sin is a queer thing [*as any fundamentalist will tell you*]. It isn't the breaking of divine commandments [but] of one's own integrity."

> "God doesn't know things. He is things."

> "My great religion is a belief in the blood, the flesh, as being wiser than the intellect. . . . What man most passionately wants

is his living wholeness and his living unison, not his own isolate salvation of his 'soul.' Man wants his physical fulfillment first and foremost, since now, once and once only, he is in the flesh and potent. . . . Whatever the unborn and the dead may know, they cannot know the beauty, the marvel of being alive in the flesh. . . . We ought to dance with rapture that we should be alive and in the flesh, and part of the living, incarnate cosmos. . . . There is nothing of me that is alone and absolute except my mind, and we shall find that the mind has no existence by itself, it is only the glitter of the sun on the surface of the waters."

Timothy Leary (1920–1996), *American psychologist, psychedelic drug missionary, and 1960s Counterculture Icon. Said of his first psychedelic drug experience (psilocybin mushrooms) that he learned more about psychology in five hours than he had in 15 years of research. Hoping to legalize LSD based on a "freedom of religion" argument, in 1966 he founded the League for Spiritual Discovery (get it?), a religion with LSD as its holy sacrament. The assistant district attorney behind the many raids and arrests at Leary's New York estate was an atheist named G. GORDON LIDDY, whom he later befriended and lectured with. President Nixon labeled Leary "the most dangerous man in America." His next-door neighbor in Folsom Prison at one point was Charles Manson. Flies on the wall were seen dropping off, dazed and confused by the conversations. Godfather of actresses Winona Ryder and Uma Thurman (daughter of his ex-wife).*

"Drugs are the religion of the 21st century."

"Pursuing the religious life today without using psychedelics drugs is like studying astronomy with the naked eye because that's how they did it in the first century AD, and besides, telescopes are unnatural."

"People use the word 'natural.'. . . What is natural to me is these botanical species which interact directly with the nervous system. What I consider artificial is four years at Harvard, and the Bible, and Saint Patrick's cathedral, and the Sunday school teachings."

"I have America surrounded."

Stanislaw J. Lec (1909–1966), *Polish-Jewish poet/aphorist.*
Holocaust survivor; escaped concentration camp in a German uniform and joined the underground in Warsaw. Author of Unkempt Thoughts *and other collections of surrealistic one-liners.*

"Sometimes the devil tempts me to believe in God."

"Perhaps God chose me to be an atheist?"

"Do I have no soul as punishment for not believing in the soul?"

"I am against using death as a punishment. I am also against using it as a reward."

"Do not ask God the way to heaven; he will show you the hardest one."

"You can change your faith without changing gods. And vice versa."

"All gods *were* immortal."

"Puritans should wear fig leaves on their eyes."

"I dreamt of a slogan for contraceptives: 'The unborn will bless you.'"

"'Thou shalt not kill' sounds like an admonition and is in fact a discovery."

"At the beginning there was the Word—at the end just the Cliché."

Richard Lederer (1938–), *American linguist and author of popular books on the word origins and oddities of English. Self-proclaimed "verbivore."*

"There once was a time when all people believed in God and the church ruled. This time was called the Dark Ages."

Gypsy Rose Lee (born Rose Louise Hovick, 1911–1970), *American burlesque entertainer. Liked to deliver intellectually stimulating patter while stripping. H. L. Mencken coined the term* ecdysiast *to convey her*

exaltedness above the common stripper. Bore a son with Otto Preminger while married to some other dude. That's the kind of morality we're dealing with here. Owned a collection of paintings by Miro, Picasso, Chagall, Max Ernst—reportedly gifts from the artists. In thanks for what? Spiritual guidance, presumably.

> "Praying is like a rocking chair—it'll give you something to do, but it won't get you anywhere."

John Lennon, MBE (1940–1980), *Liverpudlian singer/musician/composer/peace activist. Chose Yoko over the Beatles, destroying a band that might have continued producing brilliant music for decades. Changed his middle name from Winston to Ono. Strange man.*

The gospel according to John, 1966: "Christianity will go. It will vanish and shrink. . . . I don't know what will go first, rock 'n' roll or Christianity. We're more popular than Jesus now. Jesus was all right but his disciples were thick and ordinary. It's them twisting it that ruins it for me." *The Vatican accepted his apology, and if it's good enough for the Vatican. . . .*

"I've never not been political, though religion tended to overshadow it in my acid days. . . . And that religion was directly the result of all that superstar shit—religion was an outlet for my repression. . . . At one time I was so much involved in the religious bullshit that I used to go around calling myself a Christian Communist, but as [*psychologist Arthur*] Janov [*inventor of Primal therapy*] says, religion is legalized madness. It was [*Primal*] therapy that stripped away all that and made me feel my own pain. . . . You are forced to realise that your pain is really yours and not the result of somebody up in the sky. . . . This therapy forced me to have done with all the God shit. . . ."

Thus: "God is a concept by which we measure our pain. . . . I don't believe in Bible . . . I don't believe in Elvis . . . I don't believe in Beatles . . . I just believe in me. . . ." *(Not even in Elvis?? Atheism can be carried too far.)*

"Imagine there's no yada-yada. . . . It's easy if you blah blah blah. . . ."

Pope Leo XIII (Vincenzo Pecci, 1810–1903), *Italian Pope (1878–1903). Trivia: Awarded a gold medal to a cocaine-laced wine called Vin Mariani, the drink that inspired Coca-Cola.*

"The death sentence is a necessary and efficacious means for the Church to attain its ends when obstinate heretics disturb the ecclesiastical order.

Social progress? Quidem ne cogitare *(don't even think about it):* "Humanity is destined to remain as it is."

"The equal toleration of all religions is the same as atheism." *Well, then, thank God we live in the atheistic United States of America.*

Joseph Lewis (1889–1968), *president of Freethinkers of America. Founder of the Freethought Press Association and Age of Reason magazine. Author of* An Atheist Manifesto; The Tyranny of God; The Bible Unmasked; Burbank, the Infidel; *and* Voltaire, the Incomparable Infidel, *among other books.*

"Religion is all profit. They have no merchandise to buy,* no commissions to pay, and no refunds to make for unsatisfactory service and results. . . . Their commodity is fear." *Their inventories are lies. . . . Their deferred tax assets are guilt and self-abasement. . . .*

"If I had the power that the New Testament narrative says that Jesus had, I would not cure one person of blindness, I would make blindness impossible; I would not cure one person of leprosy, I would abolish leprosy."

Georg Christoph Lichtenberg (1742–1799), *German physicist, aphorist, satirist. Much admired by* SCHOPENHAUER, NIETZSCHE, TOLSTOY, FREUD, WITTGENSTEIN—*major, major names. Less admired, presumably, by the parents of the poor young girls he went for (one was 13).*

---

* This was before the dramatic rise in the price of holy water that followed the creation of OHWEC in the 1970s.

"Probably no invention came more easily to man than when he thought up heaven."

"There is a sort of transcendental ventriloquy through which men can be made to believe that something which was said on earth came from heaven."

"With most people unbelief in one thing [*e.g., God?*] is founded upon blind belief in another [*science, reason?*]."

G. Gordon Liddy (1930–), *far-right-wing American radio talk show host and Republican political strategist. Co-masterminded the Watergate break-in. Previously an FBI agent and a lawyer. Has reportedly regressed back to Roman Catholicism. From back when he had* one *thing right:*

"Looking at the Roman Catholic faith, I was unable to distinguish its assertion of a virgin birth, a return from the dead, and the bodily assumption into somewhere outside the universe of a man and his mother from among similar pagan superstitions. My last fear, the fear of God, died with my faith. I was now alone and would have to live life armed only with my own inner resources. I felt a surge of confidence and resolve like that I had experienced years before when I conquered my fear of lightning [*by climbing a tree during a thunderstorm*]. I was free."
You know, an atheist can hold his hand over a candle flame until the flesh burns. But don't try it at home.

Joe Lieberman (1942–), *American Orthodox Jewish politician; U.S. senator from Connecticut. Almost became vice president. It still would have been much, much better, believe me . . . even despite this:*

"The Constitution guarantees Americans freedom of Religion, not freedom from Religion."

Abraham Lincoln (1809–1865), *lawyer and U.S. president. What—he who said "that this nation, under God, shall have a new birth of etcetera"? Yes, him. Before he had politics to worry about, he wrote a trea- tise against Christianity, arguing that the Bible was not God's revelation and*

Jesus was not the son of God. After his assassination, his former law partner said Abraham was "an avowed and open infidel, sometimes bordering on atheism. . . . He went further against Christian beliefs and doctrines and principles than any man I ever heard."

"It will not do to investigate the subject of religion too closely, as it is apt to lead to infidelity."

"In great contests, each party claims to act in accordance with the will of God. Both may be, and one must be, wrong."

"I am approached . . . by religious men who are certain they represent the Divine Will. . . . If God would reveal his will to others, on a point so connected to my duty, it might be supposed he would reveal it directly to me."

Benjamin Barr Lindsey (1869–1943), *American judge and social reformer in Colorado. Oversaw reforms there that created juvenile courts and reduced election fraud; led in the movement to abolish child labor.*

"The churches used to win their arguments against atheism, agnosticism, and other burning issues by burning the ismists, which is fine proof that there is a devil but hardly evidence that there is a God."

Walter Lippmann (1889–1974), *American journalist, commentator, and confidant of several presidents. Appointed secretary of war during WWI by Woodrow Wilson, whom Lippmann persuaded to create the League of Nations. Socialist-turned-conservative; Jew-turned-secular humanist.*

"As long as all evils are believed somehow to fit into a divine, if mysterious, plan, the effort to eradicate them must seem on the whole futile, and even impious. . . . It is still felt, I believe, in many quarters, even in medical circles, that to mitigate the labor pains in childbirth is to blaspheme against the commandment that in pain children shall be brought forth."

"[Humankind], having ceased to believe without ceasing to be credulous, hangs, as it were, between heaven and earth and is at rest nowhere."

"It is to a morality of humanism that men must turn when the ancient order of things dissolves. When they find that they no longer believe seriously and deeply that they are governed from heaven, there is anarchy in their souls until by conscious effort they find ways of governing themselves."

John Locke *(1632–1704), English liberal political philosopher. Responsible for the creation of the United States and stuff. (Strong influence on the Founding Fathers.) A major investor in the English slave trade, despite opposing aristocracy and slavery in his writings.*

"I find every sect, as far as reason will help them, make use of it gladly: and where it fails them, they cry out, It is a matter of faith, and above reason."

Louis XIV *(1638–1715), French king. Spent every last sou in the treasury on wars, his Palace of Versailles, amazing wigs—no, I mean, amazing—and otherwise exalting His own Most Catholic Majesty. Restored religious intolerance to its former glory. Who can forget the day he revoked the Edict of Nantes?*

Upon hearing of a defeat in battle:"Has God forgotten all I have done for Him?"

H. P. Lovecraft *(1890–1937), American fantasy, horror, and science-fiction author. STEPHEN KING called him the greatest master of the classic horror tale. Before age five, announced he no longer believed in Santa Claus. "Further thought convinced him that arguments for the existence of God suffered the same weaknesses," a biographer wrote. At Sunday school, "when the feeding of Christian martyrs to the lions came up, Lovecraft shocked the class by gleefully taking the side of the lions."*

"If religion were true, its followers would not try to bludgeon their young into an artificial conformity; but would merely insist on their unbending quest for truth."

"To the scientist the joy in pursuing truth nearly counteracts the depressing revelations of truth."

"Science, already oppressive with its shocking revelations, will perhaps be the ultimate exterminator of our human species . . . for its reserve of unguessed horrors could never be borne by mortal brains if loosed upon the world."

"The world is indeed comic, but the joke is on mankind."

James Lovelock (1919–), *English Earth scientist. Father and popularizer of the Gaia hypothesis—named after the ancient Greek Mother Earth goddess—whereby the Earth functions as a single self-regulating system or "superorganism" which, perhaps consciously, manipulates the climate in order to make conditions more conducive to life, i.e., to itself. Herself. Great big load of crap? Many scientists (e.g., RICHARD DAWKINS) think so. Lovelock proclaimed in 2004 that only a switch to nuclear power could now halt global warming, which otherwise, he believes, spells d-o-o-m: mass human extinction.*

"I'm a scientist, not a theologian. I don't know if there is a God or not. Religion requires certainty. Revere and respect Gaia. Have trust in Gaia. But not faith." *"Thank Gaia." "Gaia bless you." "Oh my Gaia." Could work.*

St. Ignatius Loyola (1491–1556), *Spanish founder of the Jesuit order. His dictum: "Dei sacrificium intellectus"—the sacrifice of reason to God.*

"We should always be disposed to believe that that which appears to us to be white is really black, if the heirarchy of the Church so decides."

Lucretius (Titus Lucretius Carus, c. 94–49 B.C.E.), *Roman poet and EPICUREAN philosopher.*

"All religions are equally sublime to the ignorant, useful to the politician, and ridiculous to the philosopher."

"Even if I knew nothing of the atoms, I would venture to assert . . . that the universe was certainly not created for us by divine power."

"Nature is seen to do all things spontaneously of herself, without meddling by the gods."

"Mind cannot arise alone without body, or apart from sinews and blood. . . . You must admit, therefore, that when the body has perished, there is an end also of the spirit diffused through it is surely lunacy to couple a mortal object with an eternal one."

Sir Arnold Lunn (1888–1974), *famous skier, mountaineer, and writer. Introduced to skiing by his father, a Methodist pastor. Invented the slalom race. Eventually became an ardent Catholic. Talk about going downhill. (Did you really think we'd avoid that?)*

"Preaching is heady wine. It is pleasant to tell people where they get off."

Martin Luther (1483–1546), *German priest, theologian, and undisputed king of the Protestant Reformation. His book* On the Jews and Their Lies *proposed burning the Christ-killers' homes, synagogues, and schools confiscating their money (always popular) and curtailing their rights and liberties. Cited as an authority on Jewish matters by the Nazis.*

"Reason should be destroyed in all Christians."

John Lydon, a.k.a. Johnny Rotten (1956–), *lead "singer" with British punk-rock band the Sex Pistols. Refused in 2006 to be inducted into the Rock and Roll Hall of Fame. Appearing in a live broadcast of a British reality show in 2004, he called the show's viewers "fucking cunts," to prove he still had it, presumably.*

"Where is God? I see no evidence of God. God is probably Barry Manilow." *(This has been confirmed—Ed.)*

SECULAR * INFIDEL * NONBELIEVER * HUMANIST * RATIONALIST * FREETHINKER * AGNOSTIC * GODLESS * HERETIC * ATHEIST

M

Abu'l-ala-al-Ma'arri (short for Abu 'alaa' Ahmed ibn Abd Allah ibn Sullaiman al-Tanookhy al-Ma'arri, 973–1057 C.E.), *Syrian poet/philosopher. An eleventh-century Arab atheist! Lost his eyesight at age five. Gained his philosophical vision a little later. Dante's* Divine Comedy *was inspired by al-Ma'arri's book* Resalt Alghufran.

"Religion is a fable invented by the ancients, worthless except for those who exploit the credulous masses . . . ay, the lonesome world will always want the latest fairytales."

"The black stone [*the sacred Kaaba in Mecca*] is only a remnant of idols and altarstones." *Nonsense. It's the remnant of a UFO that crashed on earth 44,000 years ago. (Actually, it is apparently meteoric.)*

Thomas Babington Macaulay (1800–1859), *English writer, historian, and official of the Supreme Council of India, which he convinced to close Sanskrit and Arabic language schools in favor of English-only schools, resulting in large-scale outsourcing of tech support and telemarketing services from the U.S. to India. The term Macaulay's Children became a derogatory reference to Indians who adopt Western ways.*

"The Puritan hated bear-baiting, not because it gave pain to the bear, but because it gave pleasure to the spectators."

"Not two hundred men in London believe in the Bible." *And they wonder why the British Empire fell.*

Niccolo Machiavelli *(1469–1527), Italian statesman and political theorist whose ice-cold realism and naked, amoral disregard for anything except power continue even today to shock idealists and moralists and to inspire the likes of Karl Rove.*

"Our religion places the supreme happiness in humility, lowliness . . . to enable us to suffer. . . . These principles seem to me to have made men feeble . . . and easy prey to evil-minded men [*like Machiavelli's model ruler, Cesare Borgia*], who can control them more securely. . . ."

"All armed Prophets have been victorious, and all unarmed Prophets have been destroyed."

Ben Mack*, author of a* ROBERT ANTON WILSON*–ish cult novel titled* Poker without Cards: A Consciousness Thriller, *consisting of conversations between a psychiatrist and one Howard Campbell (also the name of a well-known* KURT VONNEGUT *character) about R.* BUCKMINSTER FULLER, *sci-fi, the occult, memes, religion, conspiracies, religious conspiracies. . . . Devotees spend absurd amounts of time debating whether Ben Mack and Howard Campbell exist and/or are the same person. "Campbell":*

"When these preachers on TV say God spoke to them, what the fuck? Shouldn't this be front-page news? Either God is speaking to them and we have a modern day prophet and the newfound words should be published everywhere, or they are criminals for swindling their donations."

Archibald MacLeish *(1892–1982), American Modernist poet and dramatist. Three-time Pulitzer Prize winner. His involvement with anti-Fascist organizations and his friendship with left-wing writers brought him under attack by conservative swine, including J. Edgar Hoover and Joseph McCarthy.*

"Piety's hard enough to take among the poor who have to practice it. A rich man's piety stinks. It's insufferable."

"A man who lives, not by what he loves but what he hates, is a sick man." *Red face at morning, atheists take warning!*

MAD *magazine (1952–), American satirical magazine. Responsible for destroying the respect for authority of at least one generation of adolescent boys (typically) and replacing it with smart-ass sarcasm. Its founder and publisher, William M. Gaines (d. 1992), habitually swore "On my honor as an atheist. . . ." From a 1995 issue:*

"You're a group of Christian-based, conservative organizations with several million dollars to spend. Do you: feed the hungry? Clothe the poor? Don't be so naive! You blow the millions on a series of slickly-worded, logic-bending ads espousing a widely-discredited theory that one can be 'cured' of homosexuality through counseling and prayer."

James Madison *(1751–1836), fourth U.S. president. Atheist? Deist? Well, no Christian. He and JEFFERSON defeated a bill in 1784 that would have given tax money to churches. Their bill, the Religious Freedom Act, affirmed the separation of church and state, later written into the Bill of Rights as the First Amendment. As president, Madison criticized the disgusting spectacle of government-employed chaplains praying at sessions of Congress. (The House and Senate today employ chaplains at a cost to taxpayers of $500,000 a year.)*

"Religious bondage shackles and debilitates the mind and unfits it for every noble enterprise."

On faith-based initiatives:

"The appropriation of funds of the United States for the use and support of religious societies, [is] contrary to the article of the Constitution which declares that 'Congress shall make no law respecting a religious establishment.'"

Ferdinand Magellan (1480–1521), *Portuguese explorer. The first to circumnavigate the Earth, to sail westward from Europe to Asia, and to cross the ocean he named—without any focus groups or marketing studies—the Pacific.*

> "The church says the earth is flat, but I know that it is round, for I have seen the shadow on the moon, and I have more faith in a shadow than in the church."

Bill Maher (1956–), *American comedian, political commentator, and talk show host on HBO, formerly on Comedy Central and ABC. Self-described libertarian. (Conservative-commentator-Jonah-Goldberg-described "libertine socialist.") Jewish mother, Roman Catholic father. Joked that he would bring a lawyer to confession: "Forgive me, Father, for I have sinned . . . I think you know Mr. Cohen. . . ."*

> "I hate religion. I think it's a neurological disorder."

> "What they're fighting about in the Middle East. . . . These myths, these silly little stories. . . . They take over this little space in Jerusalem where one guy flew up to heaven—no, no, this guy performed a sacrifice here a thousand million years ago. It's like, 'Who cares? What does that have to do with spirituality?'"

> *The problem with organized religion:* "You can't talk directly to god. That is bad. First you've got to talk to a priest. Then Mary. Then Jesus. . . . It's like going to the DMV."

> "Athletes: Jesus doesn't care who wins the game. So stop bothering him. I've never heard a team blame Jesus when they lose. . . ."

> "If God chose George Bush of all the people in the world, how good is God?"

> *On rappers wearing big, jewel-encrusted crucifixes:* "Isn't that what Jesus was all about? He's hanging there—'I hope my death will allow rappers to signal to chicks that they're rich so they can get laid more.'"

Norman Mailer (1923–), *American-Jewish novelist, journalist, playwright, blowhard. Ran for mayor of New York in 1969, proposing the city become a 51st state. Married six times.* WOODY ALLEN's *time traveler in* Sleeper, *describing the twentieth century to a scientist in the future, said, "This is a picture of Norman Mailer. He left his ego to the Harvard Medical School." Oh no—he'll take it with him. From* Advertisements for Myself, *1959:*

> "God like us suffers the ambition to make a destiny more extraordinary than was conceived for Him, yes God is like Me, only more so."

On William Thackeray's comparison of the omniscient third-person narrator to God:

> "God can write in the third person only so long as He understands His world. But if the world becomes contradictory or incomprehensible to Him, then God begins to grow concerned with His own nature. It's either that, or borrow notions from other Gods."

Kenan Malik, *Indian-born British writer. Author of several books on the philosophy of biology and contemporary theories of multiculturalism, pluralism and race. Ran for Parliament in the 1987 British elections as a candidate of the Revolutionary Communist Party. Now contributes articles such as "The Islamophobia Myth," "All Cultures Are Not Equal," and "Protect The Freedom to Shock"—views more usual on the right—to publications like the* Independent, Financial Times, *and* Sunday Times.

> "A decade ago, the *Independent* asked me to write an essay on TOM PAINE [for] the 200th anniversary of his great polemic, *The Age of Reason.* I began the article with a quote from SALMAN RUSHDIE's *The Satanic Verses* to show the continuing relevance of Paine's battle against religious authority. The quote was cut out because it was deemed too offensive to Muslims. The irony of censoring an essay in celebration of freethinking seemed to elude the editor."

> "Far from censoring offensive speech, a vibrant and diverse society should encourage it. . . . For it is the heretics who take

society forward. . . . Every scientific or social advance worth having began by outraging the conventions of its time."

André Malraux (1901–1976), *French author (*Man's Fate*), adventurer, and statesman. Served as a pilot for the Republican forces during the Spanish Civil War. Fought in the French Resistance in WWII; survived capture by the Gestapo. Jacqueline Kennedy, a great admirer, held a White House dinner in his honor, conversing with him entirely in French. But Barbara or Laura Bush would have done no less.*

"To the absurd myths of God and an immortal soul, the modern world in its radical impotence has only succeeded in opposing the ridiculous myths of science and progress."

Lucy Mangan, *British columnist. January 2005:*

"Now that the season of goodwill has passed, let's make a plea for greater intolerance (carefully directed). . . . The next time a woman (and it is always a woman—men have many flaws but at least they prefer to seek the answers to their problems in *Top Gear* [*the BBC car show hosted by* JEREMY CLARKSON] and Abi Titmuss [*a busty British model and porn channel hostess*] rather than the waxings and wanings of the moon) asks you what star sign you are, swears by essential oils, magnet therapy or talks about realigning anything but shelves, make a stand. Back her into a corner and talk at her about GALILEO, DARWIN, EINSTEIN, CRICK and WATSON . . . until she admits the error of her ways. For astrology and the rest to flourish it is only necessary that those with an IQ in double figures do nothing."

Mangasar Magurditch Mangasarian (1859–1943), *Turkish-born Armenian-American Protestant minister turned radical nonbeliever. Ordained a Congregationalist minister in Constantinople. After three years as a Presbyterian minister in Philadelphia, left his church and spent several years wandering in the proverbial wilderness. Reemerged in 1892 as head of the Chicago branch of the secularist Ethical Culture Society. Founded a*

rationalist group, the Independent Religious Society, where from 1900 to 1917 his lectures attracted large audiences. His books The New Catechism, The Truth about Jesus—Is He a Myth? *(answer: Yes) and* The Bible Unveiled *tore to pieces every basic tenet of Christianity. It wasn't pretty.*

From his farewell sermon to his Presbyterian congregation:

> "Your creed says that mankind is born and lives under the curse of God. . . . Your creed shows me a heaven thinly settled, a hell [well] peopled. . . . Your creed tells me that under the eternal law of predestination nothing can change the number of souls ransomed. This is fatalism. What need, then, of preaching the gospel?"

This, he suggested, should be printed on the flyleaf of the King James Bible:

> "A Collection of Writings of Unknown Date and Authorship Rendered into English From Supposed Copies of Supposed Originals Unfortunately Lost."

Irshad Manji *(1968–) Ugandan-born Muslim-Canadian author/journalist; critic of political and religious oppression and of treatment of women and homosexuals (she is both) in Muslim countries. Has expressed admiration of Israel (worse heresy than denying the existence of Allah). Instead of jihad, she advocates ijtihad, "the Muslim tradition of independent thinking." Best-selling author of* The Trouble with Islam. *Winner of Oprah Winfrey's first annual Chutzpah Award for "audacity, nerve, boldness and conviction." Described by the* New York Times *as "Osama bin Laden's worst nightmare." Alas, he and his kind remain hers. Like her friend* SALMAN RUSHDIE, *she has received numerous death threats. Lives behind bulletproof windows.*

> "Every faith has its share of literalists. . . . But only within Islam is literalism fast becoming mainstream. We Muslims, even here in the West, are routinely raised to believe that the Koran is the final and therefore perfect manifesto of God's will. . . . This is dangerous, because when abuse happens under the banner of my faith as it is today, most Muslims have no clue how to debate, dissent, revise or reform . . . because we have never been introduced to the possibility, let alone the virtue, of asking questions. . . ."

"Islam has a teaching against 'excessive laughter.' I'm not joking. [*Allah forbid.*] But does this mean that we should cry 'blasphemy' over less-than-flattering depictions of the prophet Muhammad? . . . When Muslims put the prophet on a pedestal, we're engaging in idolatry of our own. . . . Humility requires people of faith to mock themselves—and each other—every once in a while." *(People of non-faith too.)*

Marilyn Manson *(née* Brian Warner, 1969–), *American Goth-rock recording artist. Rumored to be a priest of the Church of Satan (the LAVEYans—not, Satan forbid, the Temple of Set set!). Said on MTV he wanted to be known as the person who brought an end to Christianity.*

"Christianity has given us an image of death and sexuality that we have based our culture around. A half-naked dead man hangs in most homes and around our necks. . . . Is it a symbol of hope or hopelessness?"

About Catholics trying to ban one of his concerts:

"If they think that an artist can destroy their faith, then their faith is rather fragile."

Christopher Marlowe (1564–1593), *English dramatist and poet. Marlowe (in whose play* Tamburlaine *the hero burns a Koran on stage—or so* Newsweek *reported) was quoted by a contemporary as saying, "There is no God." Was arrested for "persuading men to atheism," denying the deity of Jesus, and "jesting at the divine scriptures." Released on bail; stabbed to death in a tavern brawl before he could be tried and burned. "Christianity was enforced by law in Elizabethan England. Church attendance was mandatory. . . ."[3] Elizabeth executed three hundred Catholics. Her predecessor on the throne, her half-sister "Bloody Mary," burned three hundred Protestants. Everyone happy now?*

"Religion hides many mischiefs from suspicion."

"I count religion but a childish toy and hold there is no sin but innocence [ignorance]."—*From* The Jew of Malta.

Yann Martel (1963–), *Canadian author of* Life of Pi, *winner of the 2002 Booker Prize for fiction. To write the novel, which deals with spiritual matters far above your comprehension, he spent six months in India visiting religious sites and a year reading religious texts.*

"It is not atheists who get stuck in my craw, but agnostics. Doubt is useful for a while. . . . But we must move on. To choose doubt as a philosophy of life is akin to choosing immobility as a means of transportation."

Emma Martin (1812–1851), *British freethought pamphleteer. Author of* A Few Reasons For Renouncing Christianity And Professing and Disseminating Infidel Opinions, *and* Prayer: The Food Of Priestcraft And Bane Of Common Sense.

"The person much inclined to ask God's assistance, learns to repose on the hope of its obtainment, instead of actively seeking the good desired by his own labour."

"The bended knee is not the attitude for study."

Michael Martin, *American philosopher. Author of* Atheism: A Philosophical Justification *(1990) and* The Case against Christianity *(1991)— "two books that are considered the classic defense of atheism and the classic attack on Christianity, respectively."[4] But, not content with that, he made the case through a collection of short stories titled* The Big Domino in the Sky and Other Atheistic Tales *(1996). Wonderful Christmas, Hanukkah, or Eid al-Adha gift for the kids.*

"Since experiences of God are good grounds for the existence of God, are not experiences of the absence of God good grounds for the nonexistence of God?" *(I didn't see Him just the other day.)*

"Religious experiences are like those induced by drugs, alcohol, mental illness, and sleep deprivation: They tell no uniform or coherent story, and there is no plausible theory to account for discrepancies among them."

Harriet Martineau (1802–1876), *English abolitionist, suffragist, and, some reckon, the first sociologist. Following a two-year U.S. tour, she published* Society in America, *in which she wrote "that the condition of American women differed from that of slaves only in that they were treated with more indulgence. . . . Her writings on slavery have been credited with swaying English public opinion in favor of the North in the American Civil War."[16] Her break with religion—which she said made her "the happiest woman in England"—came during a visit to the Middle East to study the great religions. Everyone should visit the Holy Land!*

> "As the astronomer rejoices in new knowledge which compels him to give up the dignity of our globe as the center, the pride, and even the final cause of the universe, so do those who have escaped from the Christian mythology enjoy their release from the superstition which fails to make them happy, fails to make them good, fails to make them wise. . . . I would not exchange my freedom from old superstition, if I were to be burned at the stake next month, for all the peace and quiet of orthodoxy. . . ."

> "I certainly had no idea how little faith Christians have in their own faith till I saw how ill their courage and temper can stand any attack on it." *Martineau was "a doubter who believed everyone doubted."[2]*

Karl Marx (1818–1883), *German-Jewish social and political philosopher / economist / troublemaker. His father, although descended from a long line of rabbis, and although a deistic admirer of Enlightenment figures such as* VOLTAIRE *and* ROUSSEAU, *converted to Lutheranism when the Prussian authorities would not allow him to continue practicing law as a Jew.*

> "Religious suffering is, at one and the same time, the expression of, and a protest against, real suffering. Religion is the sign of the oppressed creature, the heart of a heartless world, and the soul of soulless conditions. It is"—*class?* —"the opium of the people." *("In Russia religion is the opium of the people, in China opium is the religion of the people."—Edgar Snow.)*

> "The abolition of religion as the *illusory* happiness of the people is required for *real* happiness. The demand to give up the

illusions about its condition is the *demand to give up a condition that needs illusions.*"

"The imaginary flowers of religion adorn man's chains. Man must throw away the flowers and also the chains."

"Although it is developed in the crude English style, [DARWIN's *Origin of Species*] contains the basis in natural history for our view." *Which is what really bothers the creationists.*

His favorite epigram: "De omnibus disputandum"—"Everything must be doubted." *"A pity that so many of his followers forgot the pith of that saying."*—CHRISTOPHER HITCHENS.

"Despite his atheism, Marx cannot be understood without the Bible. His myth of a perfect society, surmounting history, beyond history, is in fact the Biblical myth of Paradise on Earth. . . . What is Marxism if not Messianism?"— *Playwright Eugène Ionesco.*

Jackie Mason (born Jacob Maza, 1931–), *Jewish-American comedian. Was a rabbi—a funny one, they say—until age 28, when he was called to Comedy. Described his rejection of religion as in part a rebellion against his strict Orthodox father. With regard to Mason, there are two paramount questions: (1) Does God exist? (2) Did Mason give Ed Sullivan the finger?*

"If God exists he's an idiot. That's why I don't believe in any God. Because if that's how he behaves, I don't want to know such a person."

Marilyn Mason, *Education Officer of the British Humanist Association. No relation to MARILYN MANSON.*

"The Christian churches have a poor record on physical abuse; indeed Christian conservatives are among the few remaining supporters of corporal punishment in the westernized world, probably because of their belief in original sin and the literal truth of the Bible, with its frequent references to physical punishment of children: 'Thou shalt beat him with the rod, and shalt deliver his soul from hell' (Proverbs)."

On the torture and murder in 2000 of eight-year-old, Cote-d'Ivoire-born Victoria Climbié by her great-aunt and the woman's boyfriend, who over months inflicted 128 separate injuries on the child because, they claimed, she was possessed:

> "[The torture] went undetected partly because the North London Universal Church of the Kingdom of God congregation [and pastor] believed that the eight-year-old brought to them [by the great-aunt] *was* possessed by evil spirits."

> "Abuse of the young by religious groups often takes psychological forms, and assumes that children are the property of their parents. Parental rights to ensure 'education and teaching in conformity with their own religious and philosophical convictions' (European Convention on Human Rights) always seem to override children's rights to 'freedom of expression . . . to seek, receive and impart information and ideas of all kinds . . .' (Convention on the Rights of the Child)." *The latter has been ratified by all* countries *except the U.S. and Somalia.*

Quoting RICHARD DAWKINS:

> "We'd be aghast at the branding of 'Pro-Euro children' or 'Neo-Keynesian children,' on the basis of their parents' economic opinions. We presume that children either are too young to know what they think, or if old enough might disagree with their parents. Why, then, do we accept, without a murmur, the existence and separate education of 'Catholic children,' 'Protestant children,' 'Jewish children' and 'Muslim children'?"

Armistead Maupin (1944–), *American novelist, best known for* Tales of the City. *Once worked for Senator Jesse Helms of North Carolina, Maupin's home state. "Armistead Maupin" is an anagram of "is a man I dreamt up," which he insists is only coincidence. (Nonsense—"a man I dreamt up" obviously refers to God).*

> *The Onion: Is there a God?* "No." *Anything further to say?* "What further is there to say?" *In a writer, economy like that is divine.*

Joseph McCabe (1867–1955), *tireless English atheist and women's rights campaigner. Left the Catholic priesthood at 29 for a "life of sanity" and began to write with insane prolificacy. Wrote 250 books, including such light frolics as* The Tyranny of the Clerical Gestapo; The Totalitarian Church of Rome; Horrors of the Inquisition; Christianity and Slavery; The Church: The Enemy of the Workers, *and* How the Pope of Peace Traded in Blood.

"The more man puts into God, the less he retains in himself."

"Hourly we repeat the division of time into two parts, B.C., and A.D., and millions still think that B.C. means Benighted Chaos and A.D. means Age of Delight. In history we divide time into three parts, Ancient Times, the Middle Ages, and Modern Times; and we consider the Middle Ages a period of dark and turbulent semi-barbarism lying between two phases of civilization, ancient paganism and modern paganism."

"Atheism will in this century be the common attitude of civilized people." *(1936)*

John "Uncle John" McCarthy (1927–), *American computer scientist, based at Standford. Coined the term artificial intelligence in 1955; received the Turing Award in 1971 for his contributions to the AI field.*

"An atheist doesn't have to be someone who thinks he has a proof that there can't be a god. He only has to be someone who believes that the evidence on the God question is at a similar level to the evidence on the werewolf question."

Sir Ian McKellen (1939–), *English actor and gay rights activist. When an interviewer for the* Advocate *remarked that McKellen seemed quite calm in the aftermath of 9/11, he said, referring to living in London during the Blitz: "Well, darling, you forget—I slept under a steel plate until I was four years old." His father was a lay preacher; both grandfathers were preachers.*

Asked if he thought the film The Da Vinci Code, *in which he played Sir Leigh Teabing, should have carried a disclaimer that it is a work of fiction, as some religious groups wanted:*

"I've often thought the Bible should have a disclaimer in the front saying 'This is fiction.' I mean, walking on water?"

Bill McKibben (1960–), *American writer/environmentalist. In his first book,* The End of Nature *(1990), he observed that nature no longer exists on this planet: we have altered (or destroyed) everything—climate, flora, and fauna, the very chemistry of the oceans. Active in the Methodist Church. From his article "What It Means To Be Christian In America,"* Harper's, *September 2005:*

"America is simultaneously the most professedly Christian of the developed nations and the least Christian in its behavior. . . . In 2004, as a share of our economy, we ranked second to last, after Italy, among developed countries in government foreign aid. . . . Nearly 18 percent of American children lived in poverty (compared with, say, 8 percent in Sweden). In fact, by pretty much any measure of caring . . . childhood nutrition, infant mortality, access to preschool—we come in nearly last among the rich nations, and often by a wide margin. . . . Despite the Sixth Commandment, we are, of course, the most violent rich nation on earth, with a murder rate four or five times that of our European peers. . . . Having been told to turn the other cheek, we're the only Western democracy left that executes its citizens, mostly in those states where Christianity is theoretically strongest. . . . Teenage pregnancy? We're at the top of the charts. Personal self-discipline—like, say, keeping your weight under control? Buying on credit? Running government deficits? Do you need to ask?"

Delos B. McKown, *emeritus professor and former head of the philosophy department, Auburn University; former clergyman. Author of* The Mythmaker's Magic: Behind the Illusion of "Creation Science" *(1993).*

"The Bible is a mine rich in the ore of cognitive dissonance."

Margaret Mead (1901–1978), *American anthropologist. In* Coming of Age in Samoa *(1928), Mead reported high levels of casual premarital sex among her young female informants. Decades later—after conversion to Christianity—many of the same women denied their youthful gambols. Christian guilt had successfully been inculcated. Sex had become the dirty, shameful thing God meant it to be.*

"Creationism: the theory that Rome WAS built in day."

"It is an open question whether any behavior based on fear of eternal punishment can be regarded as ethical or should be regarded as merely cowardly."

Sidney E. Mead (1904–1999), *American religious historian. Unitarian. Enjoyed pointing out that* ABRAHAM LINCOLN *never joined a church.*

"Because from the sectarian's perspective religion is an all-or-nothing matter . . . it is impossible for him to conceive of a religiously neutral civil authority. If it is not overtly 'Christian' according to his sectarian definition it perforce must be 'infidel,' 'atheist,' 'godless,' or, as the sophisticated now commonly say, 'secular.'"

Sir Peter Medawar (1915–1987), *Brazilian-born British immunologist and science writer, best known for his research on organ transplant rejection. Nobel Laureate in medicine, 1960. Titled his autobiography* Memoirs of a Thinking Radish.

"Evolutionary theory permeates and supports every branch of biological science, much as the notion of the roundness of the earth underlies all geodesy and all cosmological theories on which the shape of the earth has a bearing. Thus antievolutionism is of the same stature as flat-earthism."

"The only certain way to cause a religious belief to be held by everyone is to liquidate nonbelievers."

"I regret my disbelief in God."

Philipp Melanchthon (1497–1560) *German theologian; key leader of the Lutheran Reformation; henchman of MARTIN LUTHER. Entered the University of Heidelberg at age 12, heedless of how excessive beer drinking stunts growth. Indeed, is described as "dwarfish, misshapen, and physically weak."*[1]

> "A man often preaches his beliefs precisely when he has lost them and is looking everywhere for them, and, on such occasions, his preaching is by no means at its worst."

Herman Melville (1819–1891), *American novelist, poet, and essayist. Nearly forgotten by the time of his death—his fall from favor blamed on his great commercial and artistic failure,* Moby Dick. *The white whale has been seen as representing evil, religion, Melville's Puritan conscience, the ultimate mystery of the universe, a "moby" (immense, impressive) dick, and/or the ungovernable passions connected thereto.*

> *Call him Ishmael:* "I'll try a pagan friend [Queequeg], thought I, since Christian kindness has turned out to be hollow courtesy."

> "The reason the mass of men fear God, and *at bottom dislike Him*, is because they rather distrust His Heart, and fancy Him all brain like a watch."

> "Already we have been the nothing we dread to be."

H. L. (Henry Louis) Mencken (1880–1956), *American journalist and critic—the best known of his time—and a passionate doubter. Coined the phrase "Bible Belt." Reported on the Scopes Monkey Trial. On Mencken's advice, defense attorney CLARENCE DARROW put the prosecutor, biblical fundamentalist WILLIAM JENNINGS BRYAN, on the stand, where Darrow made a total fool of him. It was awesome.*

> "Religion is fundamentally opposed to everything I hold in veneration—courage, clear thinking, honesty, fairness, and, above all, love of the truth."

"A man full of faith is simply one who has lost (or never had) the capacity for clear and realistic thought. He is not a mere ass: he is actually ill." *(Covered by Medicare, Medicaid, and most private insurance plans.)*

"Moral certainty is always a sign of cultural inferiority. The more uncivilized the person, the surer they are that they know precisely what is right and what is wrong."

"God is the immemorial refuge of the incompetent, the helpless, the miserable. They find not only sanctuary in His arms, but also a kind of superiority, soothing to their macerated egos; He will set them above their betters."

"Puritanism: the haunting fear that someone, somewhere, may be happy."

"It is now quite lawful for a Catholic woman to avoid pregnancy by a resort to mathematics, though she is still forbidden to resort to physics or chemistry."

"Sunday school is a prison in which children do penance for the evil conscience of their parents."

"We must respect the other fellow's religion, but only in the same sense and to the same extent that we respect his theory that his wife is beautiful and his children smart."

"To sum up: 1. The cosmos is a gigantic fly-wheel making 10,000 revolutions a minute. 2. Man is a sick fly taking a dizzy ride on it. 3. Religion is the theory that the wheel was designed and set spinning to give him the ride."

Moses Mendelssohn (1729–1786), *German-Jewish thinker and writer. Received a strictly Talmudic education until age 13, when he read Maimonides'* Guide for the Perplexed *(1190 C.E.) which, by virtue of its rationalism, was banned by centuries-old Jewish law for readers under age 25. His first book, on Plato, caused a sensation throughout Europe. Became known as the German Socrates. Received the king's permission to live in Berlin, which was forbidden to Jews. Grandfather of composer Felix Mendelssohn.*

"Among all the prescriptions and ordinances of the Mosaic law, there is not a single one which says: You shall believe or not believe. They all say: You shall do or not do. . . . Nowhere does it say: Believe O Israel, and you will be blessed; do not doubt, O Israel, or this or that punishment will befall you."

"Reader! To whatever visible church, synagogue, or mosque you may belong! See if you do not find more true religion among the host of the excommunicated than among the far greater host who excommunicated him."

Jean Meslier (1678–1733), *French priest* notorieux. *Could this be why? From his will:*

"I would like, and this would be the last and most ardent of my wishes, I would like the last king to be strangled with the guts of the last priest."

Meteorite Debris, *a.k.a. Peter Kelly, Australian atheist blogger/Web site host.*

"The more rabidly mad the church the more pews will be filled. So fundy evangelists have the biggest, and most profitable, congregations. More moderate churches with some respect for humans as intelligent human beings are losing numbers. I think a big revival would happen in the churches if child sacrifice was reintroduced. The old symbolic body and blood just doesn't cut it anymore."

John Stuart Mill (1806–1873), *English philosopher and economist. Began writing for newspapers in his teens. Among his first efforts was a defense of a religious doubter who had been jailed six years for "blasphemous libel." Not a full-blooded atheist, but a good, non-God-fearing agnostic.*

"A being who can create a race of men devoid of real freedom and inevitably foredoomed to be sinners, and then punish them for being what he has made them, may be omnipotent and various other things, but he is not what the English language has always intended by the adjective holy."

"Modern morality is derived from Greek and Roman sources, not from Christianity."

Dennis Miller (1953–), *American comedian and hypersyllabic political and social commentator. More acerbic than funny, one suddenly noticed after 9/11, when he became a vociferous Bush and Iraq war supporter. Anchored* Saturday Night Live's *"Weekend Update" segment for years. Joined Fox News in 2006.*

"Now 7-Eleven has bowed to pressure from the Moral Majority to remove *Playboy* and *Penthouse* from their newsstand. I guess to be fair you have to look at it from the fundamentalist perspective . . . because what it does is it forces a certain type of literature on somebody in a public place. It would be like, uh, oh, I don't know, say like putting the Bible in everybody's hotel room, or something crazy like that."

"These televangelists say they don't favor any particular denomination, but I think we've all seen their eyes light up at tens and twenties. . . ."

"Born again?! No, I'm not. Excuse me for getting it right the *first* time!"

Sir Jonathan Miller (1934–), *British physician and theater and opera director. Cowriter, producer, and cast member of the legendary early 1960s comedy revue* Beyond the Fringe. *Wrote and presented the 2004 BBC TV series* Atheism: A Rough History of Disbelief.

"In some awful, strange, paradoxical way, atheists tend to take religion more seriously than the practitioners."

A. A. (Alan Alexander) Milne (1882–1956), *Scottish-born English playwright, novelist, and children's writer. Creator of the cuddly atheist teddy bear character Winnie-the-Pooh.* Also wrote fine grown-up stuff, but no, nobody's interested in that.*

"The Old Testament is responsible for more atheism, agnosticism, disbelief—call it what you will—than any book ever written; it has emptied more churches than all the counterattractions of cinema, motor bicycle and golf course."

---

* Made up the "atheist" part.

Marvin "Old Man" Minsky (1927–), *American cognitive and computer scientist, known as the father of artificial intelligence. Co-founder of MIT's AI laboratory. Codeveloper of the Society of Mind theory, which attempts to explain intelligence as a product of the interaction of nonintelligent parts. (Presumably, then, an interfaith group could be intelligent.)*

"What caused the universe, and why? What is the purpose of life? How can you tell which beliefs are true? How can you tell what is good? . . . These questions seem different on the surface, but all of them share one quality that makes them impossible to answer: all of them are circular! You can never find a final cause, since you must always ask one question more: 'What caused that cause?' You can never find any ultimate goal, since you're always obliged to ask, 'Then what purpose does *that* serve?' . . . I once heard W. H. Auden say, 'We are all here on earth to help others. What I can't figure out is what the others are here for.'. . . Every culture finds special ways to deal with these questions. One way is to brand them with shame and taboo . . . [to] make those questions undiscussable. . . . [Another is to] adopt specific answers to circular questions and . . . indoctrinate people with those beliefs. [Both ways] spare whole populations from wasting time in fruitless reason loops."

George Monbiot (1963–), *left-wing British journalist and environmental and political activist. As an investigative journalist abroad, has been shot at, beaten by military police, shipwrecked, stung into a coma by hornets, sentenced to life imprisonment in absentia, pronounced clinically dead from cerebral malaria, and required to visit Texas. Both of his parents are Conservative Party officials.*

"Several million [Americans] have succumbed to an extraordinary delusion. . . . Jesus will return to Earth when certain preconditions have been met. The first of these [is] the establishment of a state of Israel. . . . The legions of the antichrist will then be deployed against Israel, and their war will lead to a final showdown in the valley of Armageddon. The Jews will either burn or convert to Christianity, and the Messiah will return to Earth. . . . Before the big battle begins, all 'true believers' will be lifted out of their clothes and wafted up to heaven during an event called the Rapture. . . . The true believers are now seeking to bring all this about. . . . We can laugh at these people, but we should not dismiss them. . . . Here we have a major political constituency—representing much of the current president [George W. Bush]'s core vote—in the most powerful nation on Earth, which is actively seeking to provoke a new world war. Its members see the invasion of Iraq as a warm-up act. . . . If the president fails to start a conflagration there, his core voters don't get to sit at the right hand of God."

Ashley Montagu (born Israel Ehrenberg, 1905–1999), *English-born anthropologist. Changed his name after moving to the United States. Chaired the anthropology department at Princeton; taught and lectured at Rutgers, Harvard, NYU, and the Unversity of California. Became a well-known guest on Johnny Carson's* Tonight Show.

"Science has proof without any certainty. Creationists have certainty without any proof."

"The Good Book—one of the most remarkable euphemisms ever coined."

Michel de Montaigne (1533–1592), *French Renaissance essayist. His personal motto was "What do I know?" (or "What do I know?" depending on the reading). At age 38, Montaigne locked himself in his library and remained in almost total isolation for ten years while he wrote his* Essays. *"[It] is one of the few books scholars can confirm SHAKE-SPEARE had in his library."*[1] *Influenced thinkers from from PASCAL to NIETZSCHE, EMERSON, and DE BEAUVOIR.*

"Man is certainly stark mad; he cannot make a worm, yet he will make gods by the dozen."

"It is setting a high value upon our opinions to roast men and women alive on account of them."

"We must not mock God. Yet the best of us are not so much afraid to offend Him as to offend our neighbors, kinsmen, or rulers." (*Wish some of* them *didn't exist either.*)

"To understand via the heart is not to understand."

Michael Moorcock (1939–), *British author best known for his often satirical science fiction and science fantasy novels. Has allowed other sci-fi authors to set stories in the fictional universe he created in his Jerry Cornelius novels. His 1967 Nebula award-winning novel* Behold the Man *tells the story [SPOILER WARNING] of a time traveler with a messiah complex who travels to 28 C.E. Palestine, falls in with a group of Jews awaiting a savior, discovers that the child of Mary and Joseph of Nazareth is a mentally retarded hunchback,* and steps into the role of Christ himself, playing it to the bitter end.*

> "The sentient may perceive and love the universe, but the universe may not perceive and love the sentient. The universe sees no distinction between the multitude of creatures and elements which comprise it. All are equal. None is favored. . . . It cannot control what it creates and it cannot, it seems, be controlled by its creations (though a few might deceive themselves otherwise). Those who curse the workings of the universe curse that which is deaf. Those who strike out at those workings fight that which is inviolate. Those who shake their fists, shake their fists at blind stars."

George Moore (1852–1933), *Irish novelist and art critic. Studied art in Paris, where he met Degas, Renoir, Monet, Pissarro, Mallarmé, and* ZOLA. *Introduced the Impressionists to an English audience. Publication of several of his novels was halted because of their anticlericalism. His 1916 novel* The Brook Kerith *supposed that a nondivine Christ did not die on the cross but was nursed back to health and eventually traveled to India to learn spiritual wisdom. That was controversial.*

> "Women have never invented a religion; they are untainted with that madness, and they are not moralists."

Michael Moore (1954–), *American documentary filmmaker and author.*

> "There's a gullible side to the American people. They can be easily misled. Religion is the best device used to mislead them."
> *"Film works too," I hear some of you thinking.*

---

* How many violations of political correctness was that?

Mark Morford, *left-wing columnist for the* San Francisco Chronicle *Web site SFGate.com, which suspended him temporarily in 2001 for a column about a female teacher accused of sex with a 13-year-old boy, in which he wrote: "From my own experience as a teenage boy, we shouldn't jump on this as a bad thing in every case." "Mark has been described . . . as a Lenny Bruce character, or as a Generation X Hunter S. Thompson, or as a liberal redneck," said an editor, "but he's always close to the edge."*

"The Vatican is instructing its priests all over the world, including those in AIDS-ravaged countries in Africa and Asia, to condemn condom use. . . . From Nicaragua to Kenya and the Philippines, where AIDS is raging like wildfire, the lie is the same: The church says condoms can kill. This is nothing new. The Vatican just really, really loathes condoms. And sex. And homosexuals. And women. And anything that might inhibit procreation, or that in any way empowers people to take control over their reproduction options, or that might somehow loosen the church's viselike grip."

"God wants us, if the happily bleak and decidedly nasty interpretation of Bible verse currently extolled by the rabid evangelical mind-set now mauling the American political and social landscape is to be believed, to use up the Earth however we see fit and stomp all over this pointless ecological blob with our macho SUVs and manly tanks and badass army boots because it's all just one giant disposable sandbox o' fun anyway, right? . . . The environment does not matter because the Earth does not matter . . . all that does matter is the imminent return of the bloody Christ. . . ."

John Morley (1st Viscount Morley of Blackburn, 1838–1923), *British Liberal statesman and newspaper editor. Served as chief secretary for Ireland and secretary of state for India, yet opposed British imperialism. Wrote monographs on* VOLTAIRE, ROUSSEAU, DIDEROT, BURKE, *and* WALPOLE. *Professed agnostic.*

"Where it is a duty to worship the sun, it is pretty sure to be a crime to examine the laws of heat."

Chris Morris (1962–), *British comedy writer and radio DJ. A segment of his spoof TV news show* Brass Eye *that satirized media coverage of pedophilia received the second highest number of complaints for a British TV show (after* Jerry Springer: The Opera*).*

> "We've had this book [the Bible] analysed and it reads like the ramblings of a drugged horse. The question tonight—is God confused like his prating truth pimps, or is he dead?"

Desmond Morris (1928–), *British zoologist. Author of popular books, beginning with* The Naked Ape *(1967), which controversially examined humans from a bluntly zoological point of view. Also an artist who has contributed significantly to the British Surrealist movement. In 1957 he curated an exhibition in London of chimpanzee paintings and drawings. The film adaptation of his doctoral thesis, "The Reproductive Behaviour of the Ten-Spined Stickleback," is widely available in adult video shops. Well, it used to be. I think.*

> "There have been many arguments about the location of the immortal human soul. Could it be in the heart, in the head, or perhaps diffused throughout the whole body—an all-pervading spiritual quality unique to the human being? The answer, it seems to me as a zoologist, is obvious enough: a man's soul is located in his testicles; a woman's in her ovaries. For it is here that we find the truly immortal elements in our constitution—our genes."

Henry M. Morris (1918–2006), *American hydraulic engineer; founder of the Institute of Creation Research; "father of modern creation science."*

> *You see, it's not so much that it isn't* true—*it's that:* "Evolution is the root of atheism, of communism, nazism, behaviorism, racism, economic imperialism, militarism, libertinism, anarchism, and all manner of anti-Christian systems of belief and practice."

> "The only way we can determine the true age of the earth is for God to tell us what it is. And since He has told us, very plainly,

in the Holy Scriptures that it is several thousand years in age, and no more, that ought to settle all basic questions of terrestrial chronology."

James Morrow (1947–), *American science fiction novelist/short story writer. Best known for his "Godhead Trilogy," beginning with* Towing Jehovah *(1994), in which the corpse of God—a two-mile long, gray-bearded white male—is discovered floating in the Atlantic Ocean; the Vatican organizes a mission to secret it away and entomb it in Arctic ice. Also author of the Nebula Award–winning short story collection* Bible Stories for Adults. *Has described the "militant agnostic humanist feminist" character in one of his novels as "closest to my own sympathies."*

"'There are no atheists in foxholes' isn't an argument against atheism, it's an argument against foxholes."

"I don't like that word [atheist] because [it suggests] a negative, a void, whereas atheists of my stripe experience their attitude as something quite positive, quite nourishing. . . . I don't like that word [agnostic] either. I find it evasive. It lacks sinew. An agnostic is an atheist who has lost his nerve." *"Bright" is just going begging. (See* CULLEN MURPHY.*)*

Lance Morrow, *American journalist.* Time *magazine writer since 1965. Author of 150* Time *cover stories.*

"If you scratch any aggressive tribalism, or nationalism, you usually find beneath its surface a religious core, some older binding energy of belief or superstition . . . that is capable of transforming itself into a death-force, with the peculiar annihilating energies of belief. . . . Religious hatreds tend to be merciless and absolute."

John Mortimer (1923–), *British barrister and creator of the fictional "Rumpole of the Bailey"—the rumpled, gruff, overliterate liberal defense-lawyer hero of some two dozen novels and a TV miniseries. Rumpole's motto is "Never Plead Guilty." Let that be our policy on Judgment Day and*

toward all other religious guilt-trip scams. Mortimer describes himself as an
"atheist for Christ."

> "I believe in everything to do with Christianity except for God. I
> believe in all the Christian ethics. But God . . . I suppose the best
> thing he could be called is uncaring. Every time I've interviewed
> Cardinals, Archbishops, their answer is free will. But it's all rot.
> The Nazi guards could have free will, because they could decide
> whether to do it or not. But not the people who were put in gas
> chambers. Nobody has satisfactorily answered this question."

Bill Moyers (1934–), *American journalist and commentator. Over 30*
Emmys and virtually every other major television journalism prize. Created
and hosted PBS's liberal public affairs program Now *and the 1988 televi-*
sion series The Power of Myth, *a series of interviews with* JOSEPH CAMP-
BELL. *Former Baptist minister.*

> "The ruins were still smoldering when the reverends Pat Robertson
> and Jerry Falwell went on television to proclaim that the [9/11] ter-
> rorist attacks were God's punishment of a corrupted America. They
> said the government had adopted the agenda 'of the pagans, and
> the abortionists, and the feminists, and the gays and the lesbians'
> not to mention the ACLU and People for the American Way. . . . Just
> as God had sent the Great Flood to wipe out a corrupted world,
> now—disgusted with a decadent America—'God almighty is lifting
> his protection from us.' Critics said such comments were
> deranged. But millions of Christian fundamentalists and conserva-
> tives didn't think so. They thought Robertson and Falwell were
> being perfectly consistent with the logic of the Bible as they read
> it. . . . Not many people at the time seemed to notice that Osama
> bin Laden had also been reading his sacred book closely and literally."

Malcolm Muggeridge (1903–1990), *a.k.a. "St. Mugg": British*
author, journalist, and media personality. Agnostic until the late 1960s, when he
located Jesus and became a prominent promoter of Christianity. "Discovered"
MOTHER TERESA, *whose example helped convert him to Roman Catholicism*
when he was 79. You can do it right up to the last minute, you know.

> "The orgasm has replaced the Cross as the focus of longing and
> the image of fulfillment."

Muhammad (c. 570–632), *Arab prophet. Allah's words to Muhammad in his bestseller* The Koran:

"Slay the idolators wherever ye find them. . . ."

"Fight those who believe not in Allah . . . nor acknowledge the religion of Truth. . . ."

"Smite ye above their necks and smite all their finger-tips off them."

"O Prophet! Exhort the believers to fight. . . ."

"Islam is peace."— George W. Bush, September 17, 2001.

Cullen Murphy (1952–), *American writer/editor. Managing editor of* Atlantic *magazine, 1985–2006. Author of* The Word According to Eve: Women and the Bible in Ancient Times and Our Own.

From his 2003 article "The Path of Brighteousness," about the adoption of "bright" as a catch-all term for nonbelievers and doubters of all shades and how "the brightness crusade" has begun to resemble organized religion in certain respects:

"Already there is an element of evangelical witness: 'By their visible example,' the brights' Web site (www.the-brights.net) explains, adherents 'can help other brights to step forward.' . . . And there is the telltale denominational urge to count the saved: in a *New York Times* op-ed article the bright philosopher DANIEL DENNETT put the number of brights in America at 27 million or more (which would place them below Catholics and Baptists in membership but well above Methodists and Lutherans). . . . There is an actual category of 'secular saints.' [*Among the canonized, he lists* EINSTEIN, DARWIN, ORWELL, BONO, *and* GELDOF.] In time a bright liturgy will surely develop, perhaps starting with the adoption of an official hymn. . . . [T]houghts turn naturally to one of the great spiritual epics of our time . . . Monty Python's *Life of Brian* [and its song] 'Always Look on the Bright Side of Life.'" *(See* TERRY GILLIAM.)

Vladimir Nabokov

Vladimir Nabokov (1899–1977), *Russian-born American novelist, world-class lepidopterist, and synaesthete (not a member of an Eastern Christian sect but someone whose mind associates colors with letters of the alphabet—a trait his son inherited). Like* JOSEPH CONRAD, *a native Slavic speaker who became a master of English prose—as he demostrated by saying, "I differ from Joseph Conradically."*

> "No free man needs a God."

> "A creative writer must study carefully the work of his rivals, including the Almighty."

Shabnam Nadiya

Shabnam Nadiya, *Bangladeshi writer, poet, and translator. From her essay "Why I Remain an Atheist":*

> "[The Koran] told me that no matter how much I read, how much I knew, no matter what love and compassion for people I held in my breast, no matter my intelligence, my talents . . . I would never ever be as good as even the lowliest of men. . . . I was a field for a man to sow his seed . . . my word was not to be trusted against that of a man, I was the gateway to hell because men would desire me. . . . How could any system of belief compete with the dignity and the respect that non-belief had to offer to me?"

Maryam Namazie, *Iranian campaigner for secularism and for women's and asylum-seekers' rights. Fled to England after the 1979 Islamic Revolution. Cosigner (with SALMAN RUSHDIE, TASLIMA NASREEN, IBN WARRAQ, and eight others) of "Manifesto: Together Facing the New Totalitarianism," a declaration issued in the wake of the Muhammad cartoons controversy of 2005. Named Secularist of the Year by Britain's National Secular Society in 2005.*

"The repeated calls for an unreserved apology for publishing 'offensive' and 'insulting' caricatures of Mohammad reminds me of the apologies that should be made to me and many like me. I'd like the offended Islamists . . . to apologise; not for their backward and medieval superstitions and religious mumbo jumbo but for their imposition of these beliefs in the form of states, Islamic laws and the political Islamic movement. If any of them want to apologise for the mass murder of countless human beings in Iran and the Middle East, and more recently in Europe, for veiling and sexual apartheid, for stoning, amputations, decapitations, Islamic terrorism and for the recent brutal attack on Tehran bus workers and so on and so forth, just email me direct." *(m.namazie@ukonline.co.uk)*

Dr. Taslima Nasreen (1962–), *Bangladeshi Muslim-born atheist physician / writer / feminist / human rights activist. In 1993, following a series of newspaper columns in which she criticized Islam's treatment of women, Bangladeshi Muslim clerics offered a $5,000 bounty for her murder, and she narrowly escaped being killed by fundamentalists. In 1994 a warrant was issued for her arrest after she called for revision of the Koran. More calls for her death followed the publication of her novel* Shame, *about the suffering of a Hindu family attacked by Muslims. Has been called "the most dangerous woman in the world," "Asia's Antigone," and "the female Salman Rushdie."*

"I don't find any difference between Islam and Islamic fundamentalists. I believe religion is the root, and from the root fundamentalism grows as a poisonous stem. . . . Maybe liberal Muslims are morally decent, but they're not following Islam honestly. Fundamentalists are."

"Koranic teaching still insists that the sun moves around the earth. How can we advance when they teach things like that?"

"The Sharia [Islamic religious] laws cannot be changed. They must be thrown out, abolished. . . . Why do we need seventh-century law now?"

Jack Nelson-Pallmeyer (1951–), *American liberal, Lutheran theologian; antiwar activist; briefly a Democratic candidate for Congress in 2006; author of ten books, including* Is Religion Killing Us? Violence in the Bible and the Qur'an *(2003). His political activism began when he first witnessed urban poverty in India. "I see the signs that say, 'God Bless America,'" he said after September 11, 2001. "Why not 'God Bless This World? . . . When terror comes to us we weep. But we should always weep for the rest of the world."*

"Religiously justified violence is first and foremost a problem of 'sacred' texts and not a problem of misinterpretation of texts."

Johann Neumann, *emeritus professor, sociology of law and religion, University of Tübingen, Germany.*

"During the whole Nazi period religious education in the schools was never forbidden, but on the contrary was generously encouraged. Despite this, after the war the churches demanded, more or less as compensation, that there must be Catholic schools and religion classes, so that 'something like that' never happened again. In fact, however, one has to ask oneself if the religion classes before and during the Nazi period didn't contribute substantially to Christian support for the war as the 'work of God,' through their inculcation of the doctrine of 'obedience to the authorities.'" *(Also see* Hitler.*)*

Rev. Michael Newdow, *American attorney, medical doctor, devout atheist, ordained minister of the Universal Life Church ("free, immediate ordination for anyone"—www.ulchq.com), and founder in 1997 of the First Amendmist Church of True Science (FACTS), which advocates church-state*

separation. Best known for filing a lawsuit against his daugher's Sacramento, California, school district for including the words "under God" in recitals of the Pledge of Allegiance. In June 2006 a judge rejected his suit to remove the frightful phrase "In God We Trust" from U.S. money on the grounds that it is a secular national slogan!

> "Clearly it's not treating atheists [as] equal with people who believe in God when you say 'In God We Trust' or we are a 'nation under God.'. . . . I couldn't care less what anyone believes. I just care that our government treats everybody equally."

New York Times *(1851–), 1993 editorial on the Boy Scouts refusing membership to a boy who would not sign a religious oath:*

> "Any organization could profit from a 10-year-old member with enough strength of character to refuse to swear falsely."

Friedrich Nietzsche *(1844–1900), German philosopher. No selection of quotes can do justice to the depth of his hatred of Christianity, which he regarded as a religion of slaves, the rabble, and the weak, motivated by a thirst for revenge and hatred of their natural superiors and of all that is positive, noble, beautiful, healthy, strong. Despised all movements toward social equality. Despite the anti-Christianity, his views reveal the real heart of conservatism—which is what makes him so interesting today. His connection to Nazism is there, though far less clear than the Nazis made out; for one thing, he hated Judaism mainly as the parent of Christianity, and despised anti-Semites.*

> "Are we not straying as through an infinite nothing? Do we not feel the breath of empty space? Has it not become colder? . . . God is dead. God remains dead. And we have killed him."— *Also sprach the "madman" in* The Gay Science.

> "Why atheism nowadays? 'The father' in God is thoroughly refuted; equally so 'the judge,' 'the rewarder.' . . . He does not hear—and even if he did, he would not know how to help."

> "A casual stroll through the lunatic asylum shows that faith does not prove anything."

"Mystical explanations are considered deep. The truth is that they are not even superficial."

"God is a gross answer, an indelicacy against us thinkers—at bottom merely a gross prohibition for us: you shall not think!"

"This alone is morality: Thou shalt not know."

"The man of belief is necessarily a dependent man. . . . He does not belong to himself, but to the author of the idea he believes."

"The only excuse for God is that he doesn't exist."

"Wherever on earth the religious neurosis has appeared, we find it tied to three dangerous dietary demands: solitude, fasting and sexual abstinence."

"Out of terror, the type has been willed, cultivated and attained: the domestic animal, the herd animal . . . the Christian."

"Two great European narcotics: alcohol and Christianity."

"I call Christianity the one great curse, the one great intrinsic depravity. . . . the one immortal blemish on the human race."

"Christianity was from the beginning, essentially and fundamentally, life's nausea and disgust with life, merely concealed behind, masked by, dressed up as, faith in 'another' or 'better' life. . . . The Christian resolution to find the world ugly and bad has made the world ugly and bad."

"The last Christian died on the cross."

Sherwin Nuland (1930–), *American surgeon; professor of bioethics and medicine at Yale; National Book Award–winning author. Raised as an Orthodox Jew.*

"I call myself an observant agnostic, because I go to shul every Saturday. The rabbi knows I'm an agnostic . . . my colleagues in the shul know I'm an agnostic, but I get carried away by the emotion of the thing."

On the separation of church and hospital: "The Hippocratic Oath starts off asking for help from Apollo, but that was just a standard form of oaths. It's an expression of the separation of medicine from religion. That was the great contribution of the

Hippocratic physicians—they were the first people to say sickness has nothing to do with God, or the gods, or whatever."

"The great progress that has been made in medicine and in science has not necessarily been made by men and women who don't believe in anything supernatural but it has been made by men and women who when they are studying it refuse to believe that there is anything supernatural."

Gary Numan (born Gary Anthony James Webb, 1958–), *English singer/songwriter and electropop pioneer. A pop phenomenon in the late 1970s and early 1980s, complete with an army of fans calling themselves Numanoids. Married a member of his own fan club. (It's not incest, legally.)*

"If nature is proof of God's amazing creation then I have truly seen the light, and the light is black. Nature is genius at its most cruel and savage. No benevolent God could have come up with such an outrage."

Ebenezer Obadare, *Nigerian scholar of political science at the London School of Economics. Author of a 2004 paper titled "In Search of a Public Sphere: The Fundamentalist Challenge to Civil Society in Nigeria."*

> "I used to teach in the Nigerian University and 90 percent of my students ended up as Evangelists or pastors. Who is going into industry? Who will do the thinking?"

Madalyn Murray O'Hair (1919–1995), *American atheist activist. Our queen. Our patron saint. Our Lady of Grace. Created the modern atheist movement in America as sure as God created the world much as we see it today. All started because students in Baltimore public schools— including her son—began each day with the Lord's Prayer or a Bible reading. In 1963 O'Hair brought a lawsuit that led to two landmark Supreme Court decisions banning compulsory prayer and Bible reading in public schools. Went on to found American Atheists, the largest U.S. atheist organization. Because of her, "Evangelistic atheism no longer feels treasonous—it just seems, like O'Hair, harsh, a little coarse, and not at all mainstream."[2] In a famous 1965 interview, Playboy called her "the most hated woman in America." No doubt she was delighted.*

> "The thing I'm most proud of is that people can say, 'I am an atheist,' in the United States today, without being called a

Communist atheist, or an atheist Communist. I separated the two words."

"I feel that everyone has a right to be insane. . . . If they want religious schools, build them! [Just] do not ask for the land to be tax-free. . . . Do not ask for [government] money for teacher's salaries, or more books, or anything else. Just go ahead and do your thing. . . . Just exactly the same as if you were a nudist. Somebody doesn't get a tax break for being a Mason, or whatever they're interested in."

"The 'Virgin' Mary should get a posthumous medal for telling the biggest goddamn lie that was ever told. . . . I'm sure she played around as much as I have."

Culbert Olson (1876–1962), *Democratic governor of California (1939–1943). Has been called the most openly atheistic elected official in U.S. history. Born in Utah to a pair of Mormons. (Sounds like something in Noah's Ark.) His mother was nonetheless a suffragette and became the first female elected official in Utah. Olsen served as president of United Secularists of America from 1957 until his death. Campaigned unsuccessfully against the adoption of "In God We Trust" as the official motto of California, among many other heroic battles against the forces of darkness.*

"Well, it was a Mormon school and the principal in his sermons to the children would arouse emotionalism and the children would become so emotional that they would declare they saw angels. Of course I did not see any angels and therefore did not join in the emotionalism. . . . I was called into the principal's office."

"General Eisenhower said, with reference to a break in the weather allowing the allied invasion of Europe to proceed 'with losses far below what we anticipated,' 'if there were nothing else in my life to prove the existence of an almighty God, that did.' I see little difference in the inanity of that statement of General Elsenhower's and a declaration by a fundamentalist church leader in Chicago that he knows the world is flat because the Bible says so."

Omar ibn al-Khattāb, Omar I (c. 581–644), *second Sunni Caliph (third in succession after* MUHAMMAD, *unless you're a Shiite); captor of Egypt, Palestine, Syria, North Africa, and Armenia from the Byzantines and for Islam.*

> "Burn the libraries, for their value is in this one book [the Koran]. . . . If these writings of the Greeks agree with the word of Allah, they are useless and need not be preserved; if they disagree, they are pernicious and ought to be destroyed." *(Who did you say preserved Greek knowledge through the Dark Ages?)*

P. J. (Patrick Jake) O'Rourke (1947–), *American humorist/writer/journalist and Research Fellow at the conservative-libertarian Cato Institute; former writer-editor at* National Lampoon *and* Rolling Stone *magazines and commentator on* 60 Minutes. *A left-wing hippie during his student days who in the 1970s made a wrong (right) turn. (All because men refuse to ask for directions.)*

> "Making fun of born-again Christians is like hunting dairy cows with a high powered rifle and scope."

George Orwell (Eric Arthur Blair, 1903–1950), *English writer/journalist. Born in Bengal, India, where his father worked for the opium department of the Civil Service. ("Opium of the people" doesn't get more literal than that.) A socialist unafraid to criticize the left, particularly the Stalinist wing. A biographer calls him an atheist.* Big Brother *(1984) has been interpreted here and there as representing God. In* Animal Farm, *the black-robed figure, the raven Moses, tells the other animals about Sugar Candy Mountain, where good, faithful, obedient animals go when they die.*

> "One must choose between God and Man, and all 'radicals' and 'progressives,' from the mildest liberal to the most extreme anarchist, have in effect chosen Man."

"No doubt alcohol, tobacco, and so forth, are things that a saint must avoid, but sainthood is also a thing that human beings must avoid. . . . Many people genuinely do not wish to be saints, and it is probable that some who achieve or aspire to sainthood have never felt much temptation to be human beings."

Thomas Otway (1652–1685), *English Restoration dramatist/poet. His plays include the comedy* The Atheist. APHRA BEHN *cast him in one of her plays.*

"Well, all I say is, honest atheism for my money."

Ouida (pen name of Maria Louise Ramé, 1839–1908), *English writer, animal rights activist, and animal rescuer who at times owned as many as 30 dogs. Between walking, feeding, bathing, and nursing them she wrote more than 40 novels as well as children's books and collections of short stories and essays.*

"Christianity has made of death a terror which was unknown to the gay calmness of the pagan."

"The radical defect of Christianity is that it tried to win the world by a bribe." *But it succeeded. . . .*

Ovid (Publius Ovidius Naso, 43 B.C.E.–17 C.E.), *Roman poet. Banished by the god, emperor, and prig Augustus, evidently because of his sexy poetry, possibly also as punishment for an affair with a royal.*

"It is convenient that there be gods, and, as it is convenient, let us so believe."

"Bribes, believe me, buy both gods and men."

Robert Owen (1771–1858), *Welsh-born businessman, socialist, and reformer. Made his father-in-law's factory at New Lanark a model for improved conditions for workers and their families, combined with commercial success. Envisioned Kibbutz-like communities; founded one in New Harmony, Indiana.*

> "All the religions of the world are based on total ignorance of all the fundamental laws of humanity. . . . Fully conscious as I am of the misery which these religions have created in the human race . . . I would now, if I possessed ten thousand lives and could suffer a painful death for each, willingly thus sacrifice them to destroy this Moloch." *We'll put him down as "nonreligious."*

SECULAR * INFIDEL * NONBELIEVER * HUMANIST * RATIONALIST * FREETHINKER * AGNOSTIC * GODLESS * HERETIC * ATHEIST

Heinz Pagels (1939–1988), *American physicist. Best known for his popular science books* The Cosmic Code, Perfect Symmetry, *and* The Dreams of Reason: The Rise of the Sciences of Complexity. *Had interesting views on the Nambu-Goldstone realization of chiral symmetry breaking. Husband of theology professor Elaine Pagels, author of* The Gnostic Gospels.

> "I like to browse in occult bookshops if for no other reason than to refresh my commitment to science."

> "Our capacity for fulfillment can come only through faith and feelings. But our capacity for survival must come from reason and knowledge. [Science is] not as resilient as commerce, religion, or politics. It needs careful nurturing. [To fail at that would be] an error that might cost us our existence."

Camille Paglia ("The 'g' is silent—the only thing about her that is," said British feminist Julie Burchill) (1947–), *American literary and social critic. Ranked #20 in a survey of the "Top 100 Public Intellectuals" in the world by editors of* Foreign Policy *and* The Prospect *in 2005. Has been called the "feminist that other feminists love to hate," a "female supremacist," and (by herself) "a feminist bisexual egomaniac." Celebrates all manner of politically incorrect sexuality, from porn on down; thumbs her nose*

at the puritanical, "victim-focused," groupthinking feminist establishment, which she has compared with cults like the Moonies. Advocates legalization of drugs and prostitution and the lowering of sexual consent laws. Recalls the time she blew up the outhouse at her Girl Scout summer camp by pouring in too much lime as "symboliz[ing] everything I would do with my life and work. Excess and extravagance and explosiveness. I would be someone who would look into the latrine of culture." Says she follows the model of "funny, prankish" Hindu gurus and Zen masters.

"As an atheist, I acknowledge that religion may be socially necessary as an ethical counterweight to natural human ferocity. The primitive marauding impulse can emerge very swiftly in the alienated young."

"Although I'm an atheist who believes only in great nature, I recognize the spiritual richness and grandeur of the Roman Catholicism in which I was raised. And I despise anyone who insults the sustaining values and symbol system of so many millions of people of different races around the world. An authentically avant-garde artist today would show his or her daring by treating religion sympathetically. Anti-religious sneers are a hallmark of perpetual adolescents." *Forever young! That's our motto.*

Thomas Paine *(1737–1809), English-born American author and revolutionary leader. Theodore Roosevelt called him a "filthy little atheist." Actually a deist (see JOHN ADAMS if you still don't know); wrote the definitive text of deism,* The Age of Reason, *which used reason to establish a belief in Nature's Designer but attacked Christianity as a system of superstition that "produces fanatics" and "serves the purposes of despotism." In England, sellers of the book were jailed for blasphemy. In dating his letters, instead of "A.D.," wrote "since the fable of Christ."*

"The world is my country, all mankind are my brethren, and to do good is my religion."

"Of all the tyrannies that afflict mankind, tyranny in religion is the worst. Every other species of tyranny is limited to the world we live in, but this attempts a stride beyond the grave and seeks to pursue us into eternity."

"Of all the systems of religion that ever were invented, there is no more derogatory to the Almighty, more unedifying to man, more repugnant to reason, and more contradictory in itself than this thing called Christianity. Too absurd for belief . . . it produces only atheists and fanatics."

"The Bible is a book that has been read more and examined less than any book that ever existed."

"What is it the New Testament teaches us? To believe that the Almighty committed debauchery with a woman engaged to be married. . . ."

"That God cannot lie, is no advantage to your argument, because it is no proof that priests can not, or that the Bible does not."

"One good schoolmaster is of more use than a hundred priests."

John Pariury, *member or maybe priest of The Universal Church Triumphant of the Apathetic Agnostic (www.apatheticagnostic.com). Motto: "We don't know and we don't care."*

"No amount of philosophical argumentation will cause there to be a god. . . . Arguments are not evidence. . . . In the end, I am left with things I have evidence for, and those I don't."

"A favored argument of theists is that atheists are not aware of everything there is to know about the cosmos. . . . The flaw in this argument is that it can equally be applied to theists."

Dorothy Parker (born Dorothy Rothschild, 1893–1967), *American writer, poet, theater critic, goddess of wit, consummate and much-consummated New Yorker. Founding member of the Algonquin Round Table drinking and smartass repartee club. Became involved in left-wing politics during the 1930s; blacklisted in the 1950s. Telegram from Parker to her editor, who was pressing her for overdue work while she was on her honeymoon: "Too fucking busy, and vice versa."*

"Well, [evangelist] Aimee Semple McPherson has written a book. . . . It may be that this autobiography is set down in sincerity, frankness and simple effort. It may be, too, that the Statue of Liberty is situated in Lake Ontario." *Dubbed McPherson "Our Lady of the Loudspeaker."*

"I went to a convent [school] in New York and was fired finally for my insistence that the Immaculate Conception was a spontaneous combustion."

Rebecca Ann Parker, *American professor of theology. Ordained United Methodist and Unitarian Universalist minister. Has written about the connection between domestic violence and abuse (which she experienced as a child) and Christian theology.*

"God required his son to suffer in order to save the world. That is an image of God as a child abuser, and Jesus is imaged as the perfect victim. He accepts the abuse and does it silently. He is praised in his religious community for accepting abuse as the highest form of love. . . . If [this is] the virtue of God's son . . . how is the victim of the priest's abuse going to find a justification for raising a protest? . . . How is the church going to see the perpetrators of abuse clearly if it can't see its own conceptualization of God as abuser?"

Matthew Parris (1949–), *British journalist, commentator, and former Conservative member of Parliament (1979–1986). Irked fellow Conservatives by speaking out for gay rights. Later came out. Said there were between 30 and 40 unannounced gay members of Parliament.*

"We move among and socialize with people with an unbelievably weird credo, and never discuss it. . . . Do you, all of you, really think that from Heaven an Albanian nun [MOTHER TERESA] cured an Indian peasant of a terminal medical condition [cancer]? Do you think a piece of metal [a locket containing Mother Teresa's picture that had touched her corpse] could carry such power? . . . European Commissioners, newspaper editors and respected columnists, party leaders and prime ministers' wives believe—or apparently do—[that] a deceased nun is being promoted among us [for beatification] by God through [this] strange agency. . . . Do these men and women accept [this claim, or promulgate it] knowingly in the higher cause of winning converts? And, if so, what other lies have been promoted in the same cause . . . ?" *The woman's husband declared the miracle "a hoax." The beatification went through in 2002.*

Blaise Pascal (1623–1662), *French mathematician, scientist, and philosopher. Left science and devoted himself to philosophy and theology following a mystical experience at age 31.*

"God is an infinite sphere whose center is everywhere and circumference is nowhere." *A jealous and despotic infinite sphere.*

"Pascal's Wager" : "If God does not exist, one will lose nothing by believing in him [*wrong*], while if he does exist, one will lose everything by not believing." *Pascal's arithmetic is wrong. Millions of nonbelievers have led full, happy, sane, undeluded lives (which already suggests God doesn't exist), so even in the vanishingly unlikely event that they are now burning in hell for eternity, they can't be said to have lost everything.*

"Nothing is more dastardly than to act with bravado toward God."

Pier Paolo Pasolini (1922–1975), *Italian film director, novelist, actor, painter. . . . Avowed atheist. Went from fascist sympathies to communism during WWII. Gay or bi. Charged in 1949 with the corruption of minors and obscene acts in public places; expelled from the Communist Party (which was evidently less tolerant of such behavior than the U.S. Republican Party). His 1975 film* Salò, *based on a novel by the* MARQUIS DE SADE, *was named Most Controversial Film of all time by* Time Out's Film Guide *in 2006. Directed a film of* The Gospel According to St. Matthew, *with the Catholic Church's support, in 1964. Public misbehavior, perversion, sadism, and the Passion of Christ, all in one actor-filmmaker: Now we've heard everything.*

"If you know that I am an unbeliever, then you know me better than I do myself. I may be an unbeliever, but I am an unbeliever who has a nostalgia for a belief."

Coventry Patmore (1823–1896), *English poet and critic. Associated with the Pre-Raphaelite Brotherhood, arguably the first avant-garde movement in the arts.*

"Fortunately for themselves and for the world, nearly all men are cowards and dare not act on what they believe. Nearly all our disasters come of a few fools having the 'courage of their convictions.'"

Ron Patterson, *contributor to the* Internet Infidel Newsletter.

"So-called Scientific Creationism is really nothing more than an attempt to give credence to an ancient Hebrew myth by trying to prove that virtually all the world's biologists, geologists, and paleontologists are a bunch of incompetent buffoons."

"Could a being create the fifty billion galaxies, each with two hundred billion stars, then rejoice in the smell of burning goat flesh?"

Pope Paul IV (Giovanni Pietro Carafa, 1476–1559). *Created the Ghetto of Rome; pronounced the Jews condemned to slavery by God.*

"If my own father were a heretic, I would personally gather the wood to burn him."

Gregory S. Paul (1954–), *paleontologist and illustrator specializing in dinosaurs. The first professional to depict them as active, warmblooded, and in some cases, feathered. (Think* Jurassic Park.) *Named six dinosaurs. (Not "Murray" and "Dave" but* Acrocanthosaurus altispinax *and shit. Is it necessary to swear so much?)*

From a paper published in the Journal of Religion and Society, *2005:*

> "[Globally,] higher rates of belief in and worship of a creator correlate with higher rates of homicide, juvenile and early adult mortality, STD infection rates, teen pregnancy, and abortion. The most theistic prosperous democracy, the U.S. . . . is almost always the most dysfunctional of the developing democracies, sometimes spectacularly so, and almost always scores poorly. . . . No democracy is known to have combined strong religiosity and popular denial of evolution with high rates of societal health. . . . The more secular, pro-evolution democracies . . . feature low rates of lethal crime, juvenile-adult mortality, sex related dysfunction, and even abortion."

> "[Within the U.S.,] the strongly theistic, anti-evolution South and Midwest have markedly worse homicide, mortality, STD, youth pregnancy, marital and related problems than the Northeast where societal conditions, secularization, and acceptance of evolution approach European norms."

Sam Peckinpah (1925–1984), *American film director (*The Wild Bunch, Straw Dogs, *and other transcendent orgies of violence). Asked if he believed in God:*

> "Yes. . . . One morning in Sausalito when the fog broke, and I was not hung over, and one night in Malibu when I recognized what the speed of light was."

Emo Phillips (Philip Soltanec, 1956–), *sublimely goofy American comedian.*

> "Oh God, please bend the laws of the universe for my convenience."

> "When I was a kid, I used to pray every night for a new bicycle. Then I realized that the Lord, in his wisdom, didn't work that way. So I just stole one and asked him to forgive me."

> "I was walking across a bridge one day, and I saw a man standing on the edge, about to jump off. So I ran over and said 'Stop! don't do it!' 'Why shouldn't I?' he said. I said, 'Well, there's so much to live for!' 'Like what?' 'Well . . . are you religious

or atheist?' 'Religious.' 'Me too! Are you Christian or Buddhist?'
'Christian.' I said, 'Me too! Are you Catholic or Protestant?'
'Protestant.' 'Me too! Are you Episcopalian or Baptist?' 'Baptist!'
'Wow! Me too! Are you Baptist Church of God or Baptist Church
of the Lord?' 'Baptist Church of God!' 'Me too! Are you original
Baptist Church of God, or are you Reformed Baptist Church of
God?' 'Reformed Baptist Church of God!' 'Me too! Are you
reformed baptist Church of God, Reformation of 1879, or
Reformed Baptist Church of God, Reformation of 1915?' He said,
' Reformed Baptist Church of God, Reformation of 1915!' I said,
'Die, heretic scum,' and pushed him off."

Pablo Picasso (1881–1973), *Spanish artist.*

"God is really only another artist. He invented the giraffe, the
elephant, and the cat. He has no real style. He just keeps on
trying other things." (*Did the sky during His blue period.*)

Steven Pinker (1954–), *Canadian-born cognitive scientist and
experimental psychologist at Harvard; previously at MIT. Leading defender
(along with allies DANIEL DENNETT and RICHARD DAWKINS) of evolutionary psychology, which sees the mind, consciousness, religious belief, etc., as
products of evolution by natural selection. (See E. O. WILSON.) Author of
several bestsellers on the subject. "Stripped to its essentials," he says, "every
decision in life amounts to choosing which lottery ticket to buy." Named one
of Time magazine's 100 most influential people in the world in 2004 and
one of Prospect and Foreign Policy's 100 top public intellectuals in 2005.*

"Religious explanations [for mysteries such as consciousness
and moral judgment] are not worth knowing because they pile
equally baffling enigmas on top of the original ones. What gave
God a mind, free will, knowledge, certainty about right and
wrong? How does he infuse them into a universe that seems to
run just fine according to physical laws? How does he get
ghostly souls to interact with hard matter? And most perplexing
of all, if the world unfolds according to a wise and merciful plan,
why does it contain so much suffering? As the Yiddish
expression says, If God lived on earth, people would break his
window."

Harold Pinter (1930–), *British playwright, director, actor, and political activist. Nobel Prize in Literature, 2005. Honorary Associate of the National Secular Society. Described the George W. Bush administration as "a bunch of criminal lunatics." Retired in 2005 from writing plays to concentrate on poetry and political activism. Came out of retirement in 2006 to write a dramatic sketch inspired by his hatred of cell phones. On some issues, a man of conscience cannot remain silent. From his Becketistical play* The Dwarfs:

> "'Do you believe in God?' 'What?' 'Do you believe in God?' 'Who?' 'God.' 'God?' 'Do you believe in God?' 'Do I believe in God?' 'Yes.' 'Would you say that again?' 'Have a biscuit.'"

Robert M. Pirsig (1928–), *American author, best known for* Zen, and the Art of Motorcycle Maintenance: An Inquiry into Values *(1974)—part novel, part autobiography, part philosophical treatise. Had an IQ of 170 at age nine. At 18 it must have been near 340. Flunked out of university. Studied Eastern philosophy in India. Spent 1960–1963 in and out of mental institutions.*

> "Metaphysics is a restaurant where they give you a thirty-thousand-page menu and no food."

> "What's happening is that each year our old flat earth of conventional reason becomes less and less adequate to handle the experiences we have and this is creating widespread feelings of topsy-turviness. As a result we're getting more and more people in irrational areas of thought . . . occultism, mysticism, drug changes and the like . . . because they feel the inadequacy of classical reason to handle what they know are real experiences."

Ignots Pistachio, *raving atheist who raves on* Raving Atheist *and other online forums.*

> "[All] religions eventually die out. . . . But atheism will live on regardless of what new religion replaces the old." *Interesting point.*

"If you want to know if you're insane, ask yourself if you have an unwavering belief, one that you could never disavow no matter what. If you answered yes, then you're insane."

"Don't believe everything you think."

William Pitt the Elder (1708–1778), *British Whig prime minister, 1766–1768. His son, the Younger, also became PM. The family fortune, hence political influence, came from his grandfather's sale of a single diamond. His favorite author was* DEMOSTHENES, *who was even more popular 250 years ago than he is today.*

"We need a religion of humanity. The only true divinity is humanity."

Pope Pius IX (Giovanni Maria Mastai-Ferretti, 1792–1878), *Romish Pope (1846–1878). From his "Syllabus of Condemned Opinions" (1850–1863) (translated from the Latin by* JOSEPH MCCABE):

"Every man is free to adopt and profess any religion, which, under the guidance of reason, he believes to be true."

Remember, these are prohibited *opinions:* "The Church has no power to lay down dogmatically that the relegion of the Catholic Church is the one true religion."

Pope Pius XI (Damiano Achille Ratti, 1857–1939), *Pope from 1929. Personally blessed the Italian planes on their way to bomb Ethiopian villages during Italy's glorious 1935 war of imperial conquest. To some writer:*

"MUSSOLINI is a wonderful man. Do you hear me? A wonderful man."

Pope Pius XII (Eugenio Pacelli, 1876–1958), *Pope from 1939. Famous for his silence and passivity during the Holocaust, as documented in* Hitler's Pope, *by historian John Cornwell, a practicing Catholic who admitted he had set out to absolve Pius. (Rather a man-absolves-dog story.)*

After the war, a vocal supporter of leniency toward Germany and its allies and of amnesty for war criminals. In 2000, Pope John Paul II elevated Pius to "Venerable"—second base on the road to canonization.

"Private ownership of the means of production is ordained by God."

"One GALILEO in two thousand years is enough."

Darrell Plank, *American computer game developer and antique fruit jar collector.*

"Every Christian who tries to escape the path of a speeding bullet with fear in his eye is an example of a 'foxhole conversion' to atheism. . . . There are a hell of a lot more of those conversions than there are of atheists to Christians."

Pliny the Elder (23–79 C.), *Roman scholar. Distantly related to* WILLIAM PITT THE ELDER. *Their gods at least minded their own damn business:*

"It is ridiculous to suppose that the great head of things, whatever it may be, pays any regard to human affairs."

Edgar Allan Poe (1809–1849), *American poet and writer of macabre tales. A progenitor of detective and crime fiction as well as science fiction. His wife Virginia was 13 when they married; he was 29. Esteemed by the nineteenth-century French avant garde while ignored in America (à la Jerry Lewis or Mickey Rourke; France's way of telling us we're too dumb to recognize our own geniuses). Arthur Conan Doyle said, "Each [of Poe's detective stories] is a root from which a whole literature has developed." ALFRED HITCHCOCK said it was Poe's stories that inspired him to make suspense films.*

"No man who has ever lived knows more about the hereafter . . . than you and I; and all religion . . . is simply evolved out of chicanery, fear, greed, imagination and poetry."

Katha Pollitt (1949–), *American poet and columnist for the* Nation. *Outspoken feminist and atheist; named "Freethought Heroine" in 1995 by the Freedom From Religion Foundation. Number 74 in* 100 People Who Are Screwing Up America, *by conservative Bernard Goldberg (who is number 73 in* 101 People Who Are *Really* Screwing America, *by Jack Huberman).*

> "For me, religion is serious business—a farrago of authoritarian nonsense, misogyny and humble pie, the eternal enemy of human happiness and freedom." *"Farrago": "a confused mixture; medley; hodgepodge." Cool.*

Pontiac (Obwandiyag, c. 1718–1769), *Ottawa Indian war leader. Reputed mastermind of "Pontiac's Rebellion" against the British military occupation of the Great Lakes region.*

> "They came with a Bible and their religion, stole our land, crushed our spirit, and now tell us we should be thankful to the Lord for being saved."

Alexander Pope (1688–1744), *English poet. After attacking a number of "hacks," "scribblers," and "dunces" in his satirical poem* The Dunciad, *he never again went out without his Great Dane and a pair of loaded pistols. (The word "overkill" was not yet in wide use.)*

> "Know then thyself, presume not God to scan / The proper study of mankind is man."

> "For modes of faith, let graceless zealots fight / He can't be wrong whose life is in the right."

> "For virtue's self may too much zeal be had / The worst of madmen is a saint run mad."

Sir Karl Popper (1902–1994), *Austrian-Jewish-born British philosopher. Leading defender of liberal democracy and "the open society,"*

whose enemies he identified as those who presume to know the future course of history (Hegelians, Marxists, Nazis—and, he might have added, believers in biblical eschatology). Preached falsifiability as the criterion for distinguishing scientific theory from nonscience. In his view, science and knowledge progress in an evolutionary process of "error elimination"—not toward truth but toward more and more interesting problems. Professed agnostic.

> "The genuine rationalist does not think that he or anyone else is in possession of the truth; he is well aware that acceptance or rejection of an idea is never a purely rational matter. . . ."

> "[O]ur knowledge can be only finite, while our ignorance must necessarily be infinite."

> "Science must begin with myths, and with the criticism of myths."

> "Why do I think that we, the intellectuals, are able to help? Simply because we, the intellectuals, have done the most terrible harm for thousands of years. Mass murder in the name of an idea, a doctrine, a theory, a religion—that is all *our* doing, *our* invention . . ."

> *Hear, O liberal:* "If we extend unlimited tolerance even to those who are intolerant . . . then the tolerant will be destroyed, and tolerance with them."

Dennis Potter (1935–1994), *British dramatist, best known for his TV series* Pennies from Heaven *and* The Singing Detective. *Accused of blasphemy for his 1969 BBC TV play* Son of Man, *an alternative view of the last days of Jesus. While dying of cancer of the pancreas and liver, he named his cancer Rupert, after Rupert Murdoch.*

> "Religion to me has always been the wound, not the bandage."

Ezra Pound (1885–1972), *American-born, fascist, treasonous, anti-Semitic, racist, segregationist, brilliant modernist poet.*

> "The act of bellringing is symbolic of all proselytizing religions. It implies the pointless interference with the quiet of other people."

Terry Pratchett (1948–), *English satirical fantasy-sci-fi author, best known for his* Discworld *novels. Has been compared to* SWIFT, *Dickens, Lewis Carroll,* EVELYN WAUGH, *P. G. Wodehouse,* DOUGLAS ADAMS. *Second only to J. K. Rowling in U.K. sales in 2003. Reputed to be Britain's most shoplifted author. Has "referred to himself as a 'Victorian-style' atheist, in the sense that he rejects supernaturalism but considers himself culturally and morally Christian. . . . He is sympathetic to the religious impulse per se as one manifestation of the essentially human quest for meaning."*[4]

"I think I'm probably an atheist, but rather angry with God for not existing."

"The trouble with having an open mind, of course, is that people will insist on coming along and trying to put things in it."

"What good's a god who gives you everything you want? . . . It's the *hope* that's important. Give people jam today, and they'll just sit and eat it. Jam *tomorrow*—now, that'll keep them going for ever."

"[Pi, the ratio of circumference to diameter of a circle] ought to be three times. You'd think so, wouldn't you? But does it? No. Three point one four one and lots of other figures. There's no end to the buggers. Do you know how pissed off that makes me? . . . It tells me that the Creator used the wrong kind of circles. . . . I mean, three point five, you could respect. Or three point three. That'd look *right*."

Marcel Proust (1871–1922), *French novelist / essayist / and critic. Graham Greene called him the greatest novelist of the twentieth century, if that means anything to you at all.*

"It has even been said that the greatest praise of God lies in the negation of the atheist, who considers creation sufficiently perfect to dispense with a creator."

Paul Provenza (1957–), *American actor/comedian/filmmaker* (The Aristocrats; Everyone Poops). *Once performed a skit on Comedy Central involving a priest hitting on a 13-year-old boy in a confessional and*

the Last Supper done as a Friar's Roast (of Jesus, obviously). Called priests "men who dress in women's clothes every Sunday."

> "I point to how irrational it is to have any reverence for religion at all. We look at the ancient Greeks with their gods on a mountaintop throwing lightning bolts and say, 'Those ancient Greeks. They were so silly. So primitive and naive. Not like our religions. We have burning bushes talking to people and guys walking on water. We're . . . sophisticated.'"

Samuel P. Putnam *(1838–1896), American "freethought" apostle. Previously a Congregationalist minister. Preaching skill is, what's the word, fungible? Author of* 400 Years of Freethought. *His poem "Why Don't He Lend a Hand?" describes human suffering and ends with the lines, "The god above us faileth, / The god within is strong." (So did he stop believing in God or just become disappointed in him?)*

> "The definition given by [HERBERT] SPENCER to Agnosticism cannot be accepted by science. 'The power which the universe manifests to us is utterly inscrutable.' Science will not affirm that anything is inscrutable. To do so is suicidal. Science will never give up the eternal struggle to know. To know what—a part of things? No, but all things. . . . It is theology that talks of the 'inscrutable' [and] puts up the bars of ignorance . . . but not science."

> "The last superstition of the human mind is the superstition that religion in itself is a good thing."

SECULAR * INFIDEL * NONBELIEVER * HUMANIST * RATIONALIST * FREETHINKER * AGNOSTIC * GODLESS * HERETIC * ATHEIST

Francois Rabelais (1494–1553), *French writer and Franciscan*

monk. His increasingly humanistic studies got him traded to the Benectines for two draft picks. Then had a tonsurectomy: went civilian, practiced medicine, and wrote his four-and-a-half bawdy, positively Rabelaisian Gargantua *and Pantagruel books—mostly under pseudonyms, which didn't prevent their being banned or his being attacked for heresy and forced into hiding. Last words:*

> "I am going to seek a Great Perhaps."

Sir Sarvepalli Radhakrishnan (1888-1975), *Indian statesman and philosopher. Oxford don who became the first vice president and second president of India.*

> "It is not God that is worshipped but the group or authority that claims to speak in His name. Sin becomes disobedience to authority, not violation of integrity."

Ayn Rand (born Alissa Rosenbaum, 1905–1982), *Russian-born American philosopher and writer. Founder of a vile, pestilential philosophy called "Objectivism" which, along with her novels* The Fountainhead *and* Atlas Shrugged, *has enthralled generations of college students. Objectivism combines fierce anticommunism, libertarian/free-market extremism, and*

Nietzschian glorification of the heroic individualist, the superior man, who owes nothing to society. Declared herself an atheist at age 13. To her, a biographer wrote, religion and communism "both subordinated man to a higher power: religion to god, communism to the state." (Which is worse—a God-fearing leftist or a right-wing atheist? Most of you liberals would—like me— prefer the leftist, wouldn't you? Interesting.)

"I am an intransigent atheist, but not a militant one. This means that I am an uncompromising advocate of reason and that I am fighting *for* reason, not *against* religion."

"Faith is the equation of feeling with knowledge." *(Compare* LESLIE STEPHEN.*)*

"All their [theists'] identifications consist of negating: God is that which no human mind can know, they say . . . God is non-man, heaven is non-earth, soul is non-body, virtue is non-profit . . . knowledge is non-reason. Their definitions are not acts of defining, but of wiping out."

"For centuries, the mystics of spirit had existed by running a protection racket—by making life on earth unbearable, then charging you for consolation and relief . . . by declaring production and joy to be sins, then collecting blackmail from the sinners."

"The cross is the symbol of torture. I prefer the dollar sign, the symbol of free trade, therefore of a free mind."

James Randi a.k.a. The Amazing Randi (born Randall Zwinge, 1928–), *Canadian-born stage magician, debunker of pseudoscience, and derider of spiritual crapola. Among his many amazing feats, he exposed the tricks used by Uri Geller and by televangelist and fraudulent faith healer (forgive the redundancy) Reverend Peter Popoff, driving the man of God into bankruptcy. His James Randi Educational Foundation offers a $1 million prize to anyone who can demonstrate any paranormal or supernatural power or phenomenon. Founding member of the Committee for Scientific Investigation of Claims of the Paranormal. Won a MacArthur Foundation "Genius" award in 1986, and the first RICHARD DAWKINS Award in 2003. Statement he devised to violate the blasphemy laws of all seven states that still had them, 1995:*

"I hereby state my opinion that the notion of a god is a superstition and that there is no evidence for the existence of any god(s). Further, devils, demons, angels and saints are myths; there is no life after death, no heaven or hell; the Pope is a dangerous, bigoted, medieval dinosaur, and the Holy Ghost is a comic-book character worthy of laughter and derision. I accuse the Christian god of murder by allowing the Holocaust to take place—not to mention the 'ethnic cleansing' presently being performed by Christians in our world—and I condemn and vilify this mythical deity for encouraging racial prejudice and commanding the degradation of women."

"Ibn al-Rawandi,"

Scholar and critic of Islam, which he abandoned in 1988. Borrowed his pseudonym (to avoid being, you know, assassinated) from a famous ninth-century Muslim skeptic/heretic who rejected the authority of any scriptural or revealed religion and argued that as a source of truth, intellect always takes precedence over religious claims.

"The myth of Islamic tolerance [was] largely invented by Jews and Western freethinkers as a stick with which to beat the Catholic Church. Islam was never a religion of tolerance. . . . Islam was spread by the sword . . . [as] the Arab empire . . . it is a religion largely invented to hold that empire together and subdue native populations. An unmitigated cultural disaster parading as God's will. Religious minorities were always second-class citizens in this empire. . . . For polytheists and unbelievers there was no tolerance at all, it was conversion or death. . . . These repulsive characteristics are written into the Quran, the hadith and the sharia. . . . There is no way that Islam can reform itself and remain Islam, no way it can ever be made compatible with pluralism, free speech, critical thought and democracy."

"Islam never really encouraged science, if by science we mean 'disinterested inquiry.' What Islam always meant by 'knowledge' was religious knowledge, anything else was deemed dangerous to the faith. All the real science that occurred under Islam occurred despite the religion not because of it."

"The mealy-mouthed and apologetic character of so much Western scholarship on Islam springs from the fact that many of these scholars, were, and are, believers in Christianity. . . . They were not keen to press the non-historical and non-divine arguments too far, since they realised that such arguments could just as well be used against their own cherished beliefs."

Raymond of Aguilers, *French eyewitness of the First Crusade (1096–1099). On the Crusaders' massacre of Muslims and Jews in Jerusalem:*

"Wonderful things were to be seen. Numbers of the Saracens were beheaded. . . . Others were shot with arrows, or forced to jump from the towers; others were tortured for several days, then burned with flames. In the streets were seen piles of heads and hands and feet. It was a just and marvelous judgment of God, that this place should be filled with the blood of unbelievers."

William Winwood Reade (1838–1875), *English historian, explorer, and philosopher. Author of* The Martyrdom of Man, *"a secular history of the Western world."*

"If, indeed, there were a judgment-day, it would be for man to appear at the bar, not as a criminal, but as an accuser."

Ronald Reagan (1911–2004), *American actor and U.S. president (1981–1989); Bible and astrology buff. As president, looking forward eagerly to World War III:*

"For the first time ever, everything is in place for the Battle of Armageddon and the second coming of Christ."

Ronald Reagan Jr. (1958–), *liberal journalist, political commentator, and radio talk show host. (Did I leave something out?) Once told Charles Manson he did not believe in God after Manson began preaching a sermon during an interview in his jail cell. A model of courage for all atheists! Asked if he would like to be president:*

"I would be unelectable. I'm an atheist. As we all know, that is something people won't accept."

Lou Reed (1942–), *American musician/singer/songwriter. Seriously into Tai Chi.*

"My God is rock 'n' roll."

Red Jacket (c. 1758–1830), *Seneca Indian chief.*

Reply to a missionary: "You have got our country, but are not satisfied; you want to force your religion upon us. . . . Brother, you say there is but one way to worship and serve the Great Spirit. If there is but one religion, why do you white people differ so much about it?"

Christopher Reeve (1952–2004), *American actor. In a gossip column following his paralyzing accident titled "Christopher Reeve: Inspirational Atheist," his brother Ben said, "We're devout atheists," adding that Christopher thought praying would be hypocritical; if he didn't pray before his accident, why pray now?*

"Even though I don't personally believe in the Lord, I try to behave as though He was watching."

John E. Remsberg (1848–1919), *American educator, superintendent of public education in Kansas, and popular freethought lecturer. (I don't think Kansas is in Kansas anymore. It was once remarkably progressive.) Thanks to Remsberg we know that there are exactly 610 contradictions in the Bible (e.g., Genesis 32:30: "I have seen God face to face"; John, 1: 18: "No man hath seen God at any time"). Could the Kansas State Board of Education ever use this guy now.*

"This doctrine of forgiveness of sin is a premium on crime. 'Forgive us our sins' means 'Let us continue in our iniquity.' . . . In teaching this doctrine Christ committed a sin for which his death did not atone, and which can never be forgiven."

Joseph Ernest Renan (1823–1892), *French philosopher and historian. Best known in his lifetime as the author of the hugely popular* Life of Jesus, *which asserted that Jesus and the Bible should be subjected to the same critical scrutiny as any other historical figures and documents. The Catholic Church went papal-bullistic.*

"Experience shows, without exception, that miracles occur only in times and in countries in which miracles are believed in, and in the presence of persons who are disposed to believe them."

"Oh God, if there is a God, save my soul, if I have a soul."

Jules Renard (1864-1910), *French writer, socialist, anticlericalist,* bêcheur *(smart-ass).*

"'The Heavenly Father feedeth the fowl of the air'—and in winter He letteth them starve to death."

"I don't know if god exists, but it would be better for his reputation if he didn't."

Heidi Reynolds-Stenson (c. 1985–), *North Carolina college student. From her article "My Path to Freethought," 2nd-place winner of the Freedom from Religion Foundation's Student Essay Contest, 2004:*

"My experiences growing up as an atheist in a conservative, Protestant suburb have made me realize just how harmful state-sponsored religion can be. . . . The church, as well as our local Campus Life group, was, in a sense, aligned with my high school. The Baptist church held their functions in our gymnasium and we held some of our sports practices in theirs. Campus Life had free reign in our hallways and lunch room and sponsors mandatory assemblies during school hours. I felt constantly bombarded with Christianity and saw many of my peers lured into the church and Campus Life by promises of friendship, community, free food and fun. . . . I began to feel more and more like I did not belong in my own high school and my own town. . . . [During a mandatory] Campus Life-sponsored assembly during school hours . . . they passed out teen Bibles at the door, preached, sang songs about Jesus and asked us to be saved if we hadn't been already. . . . [A woman speaker] addressed the young women in the audience and told us to not trust Planned Parenthood (direct quote: "It's a government conspiracy to force poor women to get abortions") and that condoms didn't do anything. . . ."

Mary Riddell, *British journalist/commentator/interviewer. Columnist for the* Observer. *Winner of our New Twist on "Opium of the People" Award:*

"Like other forms of private succour, such as Valium, Horlicks or a litre tub of chocolate-chip ice-cream, religion has limited use in the public domain."

Matt Ridley (1958–), *British science writer.*

"Readers prefer mysteries to facts, which is why books about astrology, telepathy, and the Bermuda Triangle sell so well. But science need not concede mystery to the occult. It can match it or better. Mysteries like deep geological time, a boundless universe, the big bang, relativity, quantum mechanics, the double helix, natural selection, mass extinction, and chaos theory—these are richer fare than anything the occult can offer."

Jim Rigby, *pastor of St. Andrew's Presbyterian Church in Austin, Texas, which in 2006 let a professed atheist, Robert Jensen, become a member.*

"I have spiritual friends who are trying to celebrate the mystery of life, and activist friends who are trying to change the world. . . . I don't believe either option represents a complete life. Apolitical spirituality runs the danger of giving charity instead of justice, while atheistic humanism runs the danger of offering facts instead of meaning." *But spirituality runs other, worse dangers. And meaning must ultimately be based on facts. I'm just saying.*

"Inflexible beliefs on matters where one has no experience is superstition whether one is a believer or in an atheist." *Perhaps, but on matters where one has no experience—supernatural matters, for example—not believing surely makes more sense.*

"Atheism can be the naked pursuit of truth, but anti-theism is more often the adolescent joy of upsetting and mocking religious people." *So his point is what exactly?*

"Some people argue that evolution disproves religion. I would say that evolution helps us understand why religion is inevitable in human beings. Our upper brain functions are built on top of a marshy swamp of animal instincts. . . . Much of our most important processes are irrational, even more are unconscious altogether. To say we will be purely scientific and objective is an act of imaginary dissociation from the liquid core of our own being." *He's saying religion is inevitable because it grows from our primitive instincts and irrationality. I'll buy the latter proposition. But does evolution help us understand why religion is inevitable, or why outgrowing it is inevitable?*

"Hegel defined religion as putting philosophy into pictures. Strange and foreboding topics like . . . metaphysics can be taught to almost anyone if they are put in story form. While it is important not to accept these images literally, it is just as important not to reject them literally." *I say those images are only blocking the view.*

"One last irony is that early Christians were sometimes accused of being atheists. Like true Muslims and Jews, the early Christians refused to worship human images of God." *Jesus is a human image of God. God is a human image of God.*

Tom Robbins (1936–), *American countercultural-iconish novelist* (Still Life with Woodpecker, Even Cowgirls Get the Blues). *The award just went to* MARY RIDDELL—*but, an honorable mention for:*

"Religion is not merely the opium of the masses, it's the cyanide."

"Human beings were invented by water as a means of transporting itself from one place to another." *(Replace "water" with "DNA" and you're talkin'* DAWKINS.*)*

Richard Roberts (1948–), *American televangelist, son of tele-vangelist Oral Roberts, and president of Oral Roberts University.*

"Sow a seed on your Mastercard, your Visa or your American Express, and then when you do, expect God to open the windows of heaven and pour you out a blessing." *(A tax deduction, anyway . . .)*

Pat Robertson (1930–) *American televangelist-entrepreneuer. Owns or has owned the Christian Broadcasting Network (which airs his* 700 Club *program); radio stations; a stake in an oil refinery; a gold mine in Liberia (in partnership with bloody dictator Charles Tayor, whom Robertson regularly defended in his broadcasts); a diamond mine in the Congo, which used helicopters belonging to Robertson's tax-exempt humanitarian relief organization to transport mining equipment; and of course—just to add a bit of genuine snake-oil flavor—Pat Roberton's Age-Defying Protein Shake (advertised on his tax-exempt ministry's Web site).*

On Christian love: "You say you're supposed to be nice to the Episcopalians and the Presbyterians and the Methodists and this, that, and the other thing. Nonsense. I don't have to be nice to the spirit of the Antichrist."

Fierce defender of mosque-state and synagogue-state separation: "The minute you turn [the Constitution] into the hands of non-Christian people and atheistic people they can use it to destroy the very foundation of our society."

"The great builders of our nation almost to a man have been Christians, because Christians have the desire to build something. . . . The people who have come into [our] institutions [today] are primarily termites. They are into destroying institutions that have been built by Christians. . . . The termites are in charge now . . . and the time has arrived for a godly fumigation." *Zyklon B? It was first developed as a pesticide.*

"The feminist agenda is . . . about a socialist, anti-family political movement that encourages women to leave their husbands, kill their children, practice witchcraft, destroy capitalism, and become lesbians."

Richard Robinson, *British philosopher. Oxford educated. Taught at Cornell University, then returned to teach at Oxford, where he spent the rest of his long, Godforsaken life. Author of* An Atheist's Values *(1964).*

> "The theist sometimes rebukes the pleasure-seeker by saying: 'We were not put here to enjoy ourselves; man has a sterner and nobler purpose than that.' The atheist's conception of man is, however, still sterner and nobler than that of the theist. . . . According to the atheist . . . there is no one to look after us but ourselves, and we shall certainly be defeated [by eventual extinction]. . . . When we contemplate the friendless position of man in the universe, as it is right sometimes to do, our attitude should be the tragic poet's affirmation of man's ideals of behaviour. Our dignity, and our finest occupation, is to assert and maintain our own selfchosen goods . . . of beauty and truth and virtue. . . . We are brothers without a father; let us all the more for that behave brotherly to each other."

John D. Rockefeller (1839–1937), *devout Northern Baptist. In 1911, the U.S. Supreme Court held that his company, Standard Oil, originated in illegal monopoly practices (getting secret rebates from the railroads, threatening to bankrupt his rivals, then buying them out).*

> "The good Lord gave me my money."

Gene Roddenberry (1921–1991), *television writer and producer. Creator of* Star Trek, *the most successful series in TV history. His mother, a devout Baptist who held prayer meetings at home, later abandoned religion, inspired perhaps by the godlessly logical Mr. Spock.* Trek *episodes often carried moral messages, but never religious ones. Nor, in all those years of exploring the universe, did the* Enterprise *encounter any evidence of God. (But is Trekism itself a religion?)*

> *Recalling listening to a sermon at around age 14:* "It was communion time, where you eat this wafer and are supposed to be eating the body of Christ and drinking his blood. My first impression was, 'This is a bunch of cannibals they've put me down among!' *[Beam me up, fast!].* . . . It wasn't until I was beginning to do *Star Trek* that

the subject of religion arose again. . . . People were saying I would have to have a chaplain on board the *Enterprise*. I replied, 'No, we don't.'"

Take it from a screenwriter: "We must question the story logic of having an all-knowing all-powerful God, who creates faulty Humans, and then blames them for his own mistakes."

Joe Rogan (1967–), *American comedian/actor. Host of the NBC reality show* Fear Factor. *Former U.S. tae kwon do grand champion. Open, enthusiastic, and experienced proselytizer for the psychedelic drug dimethyltriptamine (DMT), which has been called "the spirit molecule."*

"I saw a documentary on STEPHEN HAWKING, where he said he had a meeting with the pope, and that the pope said to him that it's alright to explore the universe, but not to look into the origins of the big bang, for that would be questioning God's story of creation. . . . Wow. . . . Just imagine that one of the greatest minds to come along in the last few hundred years, and he's taking directions from a cult leader that wears big goofy hats."

Joel Augustus Rogers (c. 1880–1966), *Jamaican-born African-American journalist. Covered the 1935 Italian-Ethiopian war (see* POPE PIUS XI) *for New York's* Amsterdam News. *Self-taught historian of religious and pseudoscientific racism.*

"The greatest hindrance to the progress of the Negro is that same dope that was shot into him during slavery. . . . The slogan of the Negro devotee is: Take the world but give me Jesus, and the white man strikes an eager bargain with him. . . . Another fact—there are far too many Negro preachers. Religion is the most fruitful medium for exploiting this already exploited group. As I said, the majority of the sharpers, who among the whites would go into other fields, go, in this case, to the ministry."

W. T. Root, *mid-twentieth-century American psychologist at the University of Pittsburgh. In a survey of 1,916 prison inmates, Root found*

almost none were nonreligious. (According to the U.S. Federal Bureau of Prisons in 1997, 0.2 percent of inmates were atheists. The 2001 American Religious Identification Survey found that 13 percent of the U.S. population was nonreligious.)

"Indifference to religion, due to thought, strengthens character."

Ernestine Rose (1810–1892), *Polish-born American feminist/ abolitionist/atheist. Her father, a rabbi, broke with tradition and taught her to study Torah in Hebrew. Created a monster. She rejected the Bible by age 14, left home at 17, and by 24 was giving speeches against slavery and for women's rights and religious freedom. A minister in Charleston, South Carolina, forbade his congregation to listen to "this female devil." A Maine newspaper said she was "a thousand times below a prostitute."*

"We are told that religion is natural; the belief in God universal. . . . It is an interesting and demonstrable fact, that all children are Atheists, and were religion not inculcated into their minds they would remain so. Even as it is, they are great sceptics, until made sensible of the potent weapon by which religion has ever been propagated, namely, fear."

Jean Rostand (1894–1977), *French biologist, philosopher, antinuclear and antideath-penalty activist. Agnostic. Son of* Cyrano de Bergerac *playwright Edmond Rostand. (Cyrano: "What would you have me do? Find a powerful protector . . . make my knees callous, and cultivate a supple spine, wear out my belly grovelling in the dust? No, thank you! . . . But . . . to sing, to dream, to laugh, to walk, to be alone, be free, with a voice that stirs, and an eye that still can see! . . ." Think he wasn't an atheist?)*

"Kill a man, one is a murderer; kill a million, a conqueror; kill them all, a God."

Anne Royall (1796–1854), *American writer/publisher/activist. Became the first person to lobby Congress regarding church-state separation, in response to a minister's campaign for Americans to elect only Protestants.*

Wrote arguments against slavery, in defense of Native Americans, and against the missionaries swarming "like locusts" across America, extracting donations from among the poorest and most superstitious. An admirer, President Andrew Jackson, once showed up to pay a ten-dollar fine levied against her, but was beaten to it by his secretary of war.[2]

"May the arm of the first member of Congress who proposes a national religion drop powerless from his shoulder, his tongue cleave to the roof of his mouth and all the people say amen."

Letter to a friend: "What think you, Matt, of the Christian religion? Between you and I, and the bed post, I begin to think it is all a plot of the priests. I have ever marked [them] the veriest savages under the sun."

Motto of two newspapers she founded and ran: "Good works instead of long prayers."

Richard Rubenstein (1924–), *American rabbi, theologian, and writer. Credited with coining the term genocide. Wrote about the fate of religious faith* After Auschwitz, *the title of his 1966 book. Associated with the Christian existentialists' "death of God theology" (see* THOMAS ALTIZER). *"In a technical sense he maintained, based on the Kabbalah, that God had 'died' in creating the world. However, for modern Jewish culture he argued that the death of God occurred in Auschwitz. . . . The only possibility left for Jews was to become pagans or to create their own meaning."*[1] *Became a practicing Buddhist.*

"Only the terrible accusation, known and taught to every Christian in earliest childhood [*and/or by Mel Gibson*], that the Jews are the killers of Christ, can account for the depth and persistence of this supreme hatred. In a sense, the death camps were the terminal expression of Christian anti-Semitism."

Arthur Rubinstein (1887–1982), *Polish-Jewish-American pianist. Did he believe in God?*

"No. You see, what I believe in is something much greater."

Michael Ruse (1940–), *British-Canadian philosopher of science, well known for his debates with creationists, who really don't deserve the dignity of a response, but anyway.*

"It is difficult to imagine evolutionists signing a comparable statement, that they will never deviate from the literal text of Charles Darwin's *On The Origin of Species*."

Salman Rushdie (1947–), *Indian-Muslim-born British writer. His novel* The Satanic Verses *(1988) was based on a Muslim tradition that Muhammad added verses to the Koran accepting three pagan goddesses as gods, but later said the devil made him do it. The novel was banned and publically burned in Muslim countries as well as in India. Iran's leader, AYATOLLAH RUHOLLAH KHOMEINI, issued a fatwa declaring the book blasphemous and offering a $2 million reward for Rushdie's assassination. Bookstores that carried Rushdie's book were firebombed. Several people who helped translate or publish it were attacked; one was killed. Iran's spiritual leader reaffirmed the fatwah in 2005. In 1998 Rushdie declared himself an atheist, squelching ugly rumors that he had reverted to Islam. Married to Indian model and actress Padma Lakshmi. That could take most men's minds off of a fatwah. From an op-ed written weeks after 9/11:*

"'This isn't about Islam.' The world's leaders have been repeating this mantra for weeks [in order to protect innocent Muslims and maintain good relations with Muslim countries]. . . . The trouble with this necessary disclaimer is that it isn't true. Of course this is about Islam."

"The idea of the sacred is quite simply one of the most conservative notions in any culture, because it seeks to turn other ideas—uncertainty, progress, change—into crimes. . . . The Islamic Reformation has to begin here, with an acceptance that all ideas, even sacred ones, must adapt to altered realities. Broad-mindedness is related to tolerance; open-mindedness is the sibling of peace." *For this, Rushdie must die like a dog.*

John Ruskin (1819–1900), *English art critic and the most famous cultural theorist of his day. A bad review from Ruskin, and your career as an artist was, as the Victorians said, toast. (Then again: James Whistler won a libel suit against him over a review.) Gave away most of his inheritance, declaring it was impossible to be a rich socialist.*

"I know few Christians so convinced of the splendor of the rooms in their Father's house, as to be happier when their friends are called to those mansions."

"I never yet met a Christian whose heart was thoroughly set upon the world to come . . . who cared about art at all."

Bertrand Russell (1872–1970), *British mathematician and philosopher. Jailed for six months during WWI for writing an antiwar article. (A jailer asked his religion: "Agnostic," Russell replied. Said the jailer, who had never heard of that religion: "I guess we all worship the same God.") His irreligion and advocacy of sexual freedom got him barred from teaching in New York by the state Supreme Court in the early 1940s. Awarded the 1950 Nobel Prize in Literature as "the champion of humanity and freedom of thought."*

On the fallacy in the argument of the First Cause, i.e., that everything except *God must have a cause:* "If there can be anything without a cause, it may just as well be the world as God, so that there cannot be any validity in that argument. It is exactly of the same nature as the Indian's view, that the world rested upon an elephant and the elephant rested upon a tortoise; and when they said, 'How about the tortoise?' the Indian said, 'Suppose we change the subject.'" *(Or as a variation goes, "Another tortoise." And beneath that? "Another." And then? "It's tortoises all the way down.")*

"If I were granted omnipotence, and millions of years to experiment in, I should not think Man much to boast of as the final result of all my efforts."

"Where there is evidence, no one speaks of 'faith.' We do not speak of faith that two and two are four or that the earth is round."

"The most savage controversies are those about matters as to which there is no good evidence either way. Persecution is used in theology, not in arithmetic."

"The fundamental cause of trouble in the world today is that the stupid are cocksure while the intelligent are full of doubt."

"One is often told that it is a very wrong thing to attack religion, because religion makes men virtuous. So I am told; I have not noticed it . . ."

"The splendour of human life, I feel sure, is greater to those who are not dazzled by the divine radiance."

"So far as I can remember, there is not one word in the Gospels in praise of intelligence."

"Every single bit of progress in humane feeling, every improvement in the criminal law, every step toward the diminution of war, every step toward better treatment of the colored races, or every mitigation of slavery . . . has been consistently opposed by the organized churches of the world. I say quite deliberately that the Christian religion, as organized in its churches, has been and still is the principal enemy of moral progress in the world."

"The Chinese said they would bury me by the Western Lake and build a shrine to my memory. I might have become a god, which would have been very chic for an atheist."

Dora Black Russell (1894–1986), *British women's rights activist. Cofounder (with H. G. WELLS and John Maynard Keynes) of the Workers' Birth Control Group. Member of the Heretics Society. Believed both men and women were polygamous by nature—perhaps because she was the wife of BERTRAND RUSSELL.*

"When the male of the species, enamored of his stargazing, set up a God outside this planet as arbiter of all events upon it, and repudiated nature, together with sex, for a promised dream of a future life, he turned his back on that creative life and inspiration that lay within himself and his partnership with woman. In very truth he sold his birthright for a mess of potage."

SECULAR * INFIDEL * NONBELIEVER * HUMANIST * RATIONALIST * FREETHINKER * AGNOSTIC * GODLESS * HERETIC * ATHEIST

S

Marquis de Sade (Donatien Alphonse François, 1740–1814),

French sadist, libertine, and pornographer. Celebrated as a philosopher by twentieth-century intellectuals from Guillaume Apollinaire (who called Sade "the freest spirit that has yet existed") to SIMONE DE BEAUVOIR. Described himself in his will as "atheistic to the point of fanaticism." His "Dialogue Between a Priest and a Dying Man" has been described as "clearly the work of someone with contempt for religion." His favorite whipping boy, so to speak: Christianity.

> "The idea of God is the sole wrong for which I cannot forgive mankind."

> "It requires only two things to win credit for a miracle: a mountebank and a number of silly women."

Carl Sagan (1934–1996), *American astronomer and science popularizer. "My candidate for planetary ambassador . . . a beacon of clear light in a dark world of alien abductions and 'real-life X-files,' of psychic charlatans and New Age airheads, of fatcat astrologers giggling all the way to the millennium."—RICHARD DAWKINS. "Carl never wanted to believe. He wanted to know."—Sagan's wife, ANN DRUYAN.*

> "Life is but a momentary glimpse of the wonder of this

astonishing universe, and it is sad to see so many dreaming it away on spiritual fantasy."

"Modern science has been a voyage into the unknown, with a lesson in humility waiting at every stop. Many passengers would rather have stayed home."

"In 1993, the supreme religious authority of Saudi Arabia, Sheik Abdel-Aziz Ibn Baaz, issued an edict, or fatwa, declaring that the world is flat. [*In 1966 the same sage wrote, 'The Holy Koran, the Prophet's teachings, the majority of Islamic scientists, and the actual facts all prove that the sun is running in its orbit . . . and that the earth is fixed and stable.'*] Anyone of the round persuasion does not believe in God and should be punished. When the movie *Jurassic Park* was shown in Israel, it was condemned by some Orthodox rabbis because it taught that dinosaurs lived a hundred million years ago. . . . The clearest evidence of our evolution can be found in our genes, but evolution is still being fought, ironically by those whose own DNA proclaims it." *Maybe their particular DNA proclaims our lowly ancestry a bit too loudly. . . .*

. . . indeed: "A celibate clergy is an especially good idea, because it tends to suppress any hereditary propensity toward fanaticism."

"In a scientific age, what is a more reasonable and acceptable disguise for the classic religious mythos than the idea that we are being visited by messengers of a powerful, wise and benign advanced civilization?" *(Compare* THOMAS E. BULLARD.*)*

"The sacred truth of science is that there are no sacred truths."

"It often happens that scientists say, 'You know that's a really good argument; my position is mistaken,' and then they actually change their minds and you never hear that old view from them again. . . . I cannot recall the last time something like that happened in politics or religion."

Carl Sandburg (1878–1967), *American poet, biographer* (ABRAHAM LINCOLN—*Pulitzer Prize), folklorist.*

"To work hard, to live hard, to die hard, and then to go to hell after all would be too damned hard."

Margaret Sanger (1883–1966), *American birth control activist. Founder of what became Planned Parenthood. Daughter of an outspoken free-thinker; Margaret and her siblings were labeled "heathens," "heretics," and "devil's children" by the good people of their upstate New York town. Her religious awakening began at dinner one night when her father, seeing her thanking God for her bread, asked if God was a baker. Jailed eight times for publishing information about birth control and venereal disease prevention and for opening America's first birth control clinic (1916)—or as the authorities described it, "maintaining a public nuisance."*

> *Masthead of her newsletter,* The Woman Rebel: "No Gods, No Masters."

> "If Christianity turned the clock of general progress back a thousand years, it turned back the clock two thousand years for woman."

> "Cannibals at least do not hide behind the sickening smirk of the Church. . . . Their tastes are not so fastidious, so refined, so Christian, as those of our great American coal operators. . . . Remember the women and children who were sacrificed so that JOHN D. ROCKEFELLER JR., might continue his noble career of charity and philanthropy as a supporter of the Christian faith."

George Santayana (1863–1952), *Spanish-born American philosopher. "He contended that religion, although factually untrue, should be cherished as irrational poetry."*[3] *Raised by unbeliever parents yet sent to Catholic schools, as a child he believed there is no God and that the Virgin Mary is His mother.*

> "My atheism, like that of SPINOZA, is true piety towards the universe and denies only gods fashioned by men in their own image, to be servants of their human interests; and . . . even in this denial I am no rude iconoclast, but full of secret sympathy with the impulses of idolators."

> "Men become superstitious not because they had too much imagination, but because they were not aware that they had any."

> "Faith in the supernatural is a desperate wager made by man at the lowest ebb of his fortunes."

"It is pathetic to observe how lowly the motives are that religion, even the highest, attributes to the deity. . . . To be given the best morsel, to be remembered, to be praised, to be obeyed blindly. . . . No religion has ever given a picture of deity which men could have imitated without the grossest immorality."

Buddhism had tried to quiet a sick world with anesthetics; Christianity sought to purge it with fire.

"It is not worldly ecclesiatics that kindle the fires of persecution, but mystics who think they hear the voice of God."

"It is easier to make a saint out of a libertine than out of a prig."

"There is no cure for life and death save to enjoy the interval."

Jean-Paul Sartre (1905–1980), *French existentialist philosopher, novelist, and playwright. To Sartre, life has no meaning except that which we choose and dedicate our lives to. . . . There's no God to dictate right and wrong. But even if God existed, it would still be necessary to reject him, "since the idea of God negates our freedom. . . . We must see human beings as liberty incarnate."[10] A moral life is heroic only when freely chosen, and is all the more necessary in a godless world. Rejected the Nobel Prize in Literature in 1964—the only person ever to do so—saying he did not want to be turned into a cultural icon. Bit late for that, quoi.*

As good a time as any: "And then one day . . . to while away the time, I decided to think about God. 'Well,' I said, 'he doesn't exist.' It was something authentically self-evident . . . I settled the question once and for all at the age of twelve."

"DOSTOYEVSKY said, 'If God did not exist, everything would be possible [*i.e., permissible*].' That is the very starting point of existentialism."

"Existentialism . . . isn't trying to plunge man into despair at all. . . . We mean only to say that God does not exist, and that it is necessary to draw the consequences of his absence right to the end."

"Existentialism isn't so atheistic that it wears itself out showing that God doesn't exist. Rather, it declares that even if God did exist, that would change nothing."

"The existentialist . . . finds it extremely embarrassing that God does not exist, for there disappears with him all possibility of finding values in an intelligible heaven."

"There is no human nature, because there is no God to have a conception of it. Man simply is. . . . Man is nothing else but that which he makes of himself. That is the first principle of existentialism."

"If God exists, man does not exist; if man exists, God does not exist."

"Hell is other people." *("Hell is yourself."—*Tennessee Williams*)*

Dan Savage (1964–), *Homo-American; writes a syndicated sex and relationship advice column, "Savage Love"; edits the Seattle weekly paper* the Stranger. *After Republican Senator Rick Santorum compared adult homosexual sex to child molestation, incest, and bestiality in a 2003 interview, Savage printed a definition of "santorum" as "the frothy mix of lube and fecal matter that is sometimes the byproduct of anal sex" and, by "Google bombing," made that definition the first result for a Google search on "santorum." Calls himself "a wishy-washy agnostic" and a Catholic "in a cultural sense, not an eat-the-wafer, say-the-rosary, burn-down-the-women's-health-center sense." Purchased the late Ann Landers's desk.*

"John Paul II had more 'no's' for straight people than he did for gays. But when he tried to meddle in the private lives of straights, the same people who deferred to his delicate sensibilities where my rights were concerned suddenly blew [him] off. Gay blowjobs are expendable, it seems; straight ones are sacred. . . . So . . . I'm sorry the old bastard's dead. . . . But I'm not so sorry that I won't stoop to working John Paul II into a column about zombie fetishism."

"I realized if I wanted to live in a fabulous house and have sex with young men I didn't need to be a priest to do that anymore. It used to be the only way you could do that, but the world has changed now for the better."

Friedrich von Schiller (1759–1805), *German poet and dramatist. A greater poet than his good friend GOETHE, according to Beethoven (who set Schiller's Ode to Joy to music in the final movement of his Ninth Symphony).*

> "Which religion do I profess to follow? None! And why?" *Tell us.* "Because of religion."

> "A healthy nature needs no God or immortality. There must be a morality which suffices without this faith."

Arthur Schnitzler (1862–1931), *Austrian playwright and physician. The "frank" treatment of "sex" in his plays got them branded as pornography. ("Jewish filth," said HITLER.) In his play* Professor Bernhardi, *a Jewish doctor is jailed for turning away a Catholic priest so that a patient will not realize she is on the point of death. "I write of love and death," Schnitzler said. "What other subjects are there?" Kept a diary from age 17 until his death—52 years. Recorded his many, often simultaneous sexual relationships and, for some years, every orgasm.*

> "Martyrdom has always been a proof of the intensity, never of the correctness of a belief."

Arthur Schopenhauer (1788–1860), *German philosopher. Great and strangely cheerful apostle of pessimism. Denied the existence of God in any form. Didn't believe in the pursuit of truth through science, either, but rather, through art. Much interested in Hindu and Buddhist thought, which he saw as akin to his own. The brilliance of his aphorisms proclaims the glory of God's creation.*

> "That a god like Jehovah should have created this world of misery and woe, out of pure caprice, and because he enjoyed doing it, and should then have clapped his hands in praise of his own work, and declared everything to be very good—that will not do at all!"

> "Religions are like fireflies. They require darkness in order to shine."

"Faith and knowledge are related as the scales of a balance; when the one goes up, the other goes down."

"The prayer 'lead me not into temptation' means 'Let me not see who I am.'"

"Man excels all the animals even in his ability to be trained. . . . Religion in general constitutes the real masterpiece of the art of training."

"In every religion it soon comes to be the case that faith, ceremonies, rites and the like, are proclaimed to be more agreeable to the Divine will than moral actions; the former . . . gradually come to be looked upon as a substitute for the latter."

"If a public proclamation were suddenly made announcing the repeal of all the criminal laws, I fancy neither you nor I would have the courage to go home from here under the protection of religious motives. If, in the same way, all religions were declared untrue, we could, under the protection of the laws alone, go on living as before, without any special addition to our apprehensions or our measures of precaution."

"The Catholic religion is an order to obtain heaven by begging, because it would be too troublesome to earn it. The priests are the brokers for it."

"Whether one makes an idol of wood, stone, metal, or constructs it from absolute ideas, it is all the same; it is idolatry, whenever one has a personal being in view to whom one sacrifices, whom one invokes, whom one thanks."

"Philosophy lets the gods alone, and asks in turn to be let alone by them."

"If continued existence after death could be proved to be incompatible with the existence of gods . . . [believers] would soon sacrifice these gods to their own immortality, and be hot for atheism."

"To desire immortality is to desire the perpetuation of a great mistake."

Olive Schreiner (1855–1920), *South African writer, socialist, and feminist. Born in Basutoland (now Lesotho), the ninth of 12 children of Calvinist missionaries. Her first novel,* The Story of an African Farm, *written before she was 20 and published under a male pseudonym, was an immediate success and gained her entrée into radical intellectual circles in Europe. Regarded as one of the first feminist novels, it describes her own loss of faith. Her home was once burned down by whites who regarded her racial views as too liberal; to present-day readers, her attitude toward blacks seems quite patronizing enough, thank you.*

> "But we, wretched unbelievers, we bear our own burdens; we must say, 'I myself did it, *I*. Not God, not Satan; I myself!'"

Charles Schultz (1922–2000), *American cartoonist. His* Peanuts *strip ran from 1950 to 2000. Raised as a Lutheran. Taught Sunday school as a young man. In the 1965 animated cartoon* A Charlie Brown Christmas, *Linus—who Schultz said represented his spiritual side—quotes the New Testament to explain "what Christmas is all about." By the 1980s Schultz had seen the light. Nevertheless, his strips continued to be hijacked and plundered for Christian morals, as in the bestseller* The Gospel According to Peanuts. *From a 1999 interview:*

> "The term that best describes me now is 'secular humanist.' . . . I despise those shallow religious comics. *Dennis the Menace*, for instance, is the most shallow. When they show him praying—I just can't stand that sort of thing, talking to God about some cutesy thing that he'd done during the day." *The comics page— that is where the contest between faith and reason will ultimately be decided.*

Frithjof Schuon (1907–1998), *Swiss-born religious philosopher and Sufi (Muslim mystic). A founder of the "Traditionalist School," which holds that all authentic religious traditions are true, and derive from a primordial and universal something-or-other called the Primordial Tradition (a.k.a.* religio perennis *or "underlying Religion"). Unlike other cultures,*

however, the Western world has lost its connection to the PT, and as a result the world is in a state of intellectual and spiritual decline.

"The intellectual—and thereby the rational—foundation of Islam results in the average Muslim having a curious tendency to believe that non-Muslims either know that Islam is the truth and reject it out of pure obstinacy, or else are simply ignorant of it and can be converted by elementary explanations; that anyone should be able to oppose Islam with a good conscience quite exceeds the Muslims' powers of imagination. . . ."

George S. Schuyler (1895–), *African-American journalist.*
Managing editor of the Messenger, *a black-run left-wing magazine launched in 1917, whose editors the U.S. attorney general described as "the most dangerous Negroes in the United States." Reported on Jim Crow in the South in the 1920s. Contributed articles such as "Black America Begins to Doubt" (1932) to H. L.* MENCKEN's *American Mercury and "Black Man's Burden—Religion" (1935) to E.* HALDEMAN-JULIUS's *American* Rationalist. *Helped expose "the extortion, misappropriation, and adultery within Black churches."[11] A fervent anticommunist (unlike most black intellectuals of his day), he titled his autobiography* Black and Conservative *and eventually wrote for the John Birch Society magazine.*

"Practically all of the incompetents and undesirables who have been barred from other walks of life have rushed into the ministry for the exploitation of the people. . . . Almost anybody of the lowest type may go into the Negro ministry."

"On the horizon loom a growing number of iconoclasts and Atheists, young black men and women who can read, think, and ask questions, and who impertinently demand to know why Negroes should revere a God who permits them to be lynched, jim-crowed and disfranchised."

Scientific American *magazine, Editorial, April 2005:*

"In retrospect, this magazine's coverage of so-called evolution has been hideously one-sided. . . . Why were we so unwilling to suggest that dinosaurs lived 6,000 years ago or that a cataclysmic flood carved the Grand Canyon? . . . As editors, we had no business being persuaded by mountains of evidence. . . . Nor should we succumb to the easy mistake of thinking that scientists understand their fields better than, say, U.S. senators or best-selling novelists do. Indeed, if politicians or special-interest groups say things that seem untrue or misleading, our duty as journalists is to quote them without comment or contradiction. To do otherwise would be elitist and therefore wrong." *Critics say* Scientific American *has grown more sarcastic since its founding in 1845.*

John Selden (1584–1654), *English lawyer, member of Parliament, and scholar of "oriental" religions. Along with his duller books on law and civil administration, he wrote histories of dueling and of tithes. (Guess which one the bishops forced him to retract.)*

"*Scrutamini scripturas* [Let us examine the scriptures]. These two words have undone the world."

"The clergy would have us believe them against our own reason, as the woman would have her husband believe against his own eyes."

Etta Semple (1855–1914), *American feminist-freethought novelist and activist. Helped found the Kansas Freethought Association. Published a newsletter, the* Freethought Ideal.

"If, in plain word I don't want to go to heaven . . . whose business is it but my own?" *Ain't nobody's business.*

Seneca (Lucius Annaeus Seneca "the Younger," 4–65 A.D.), *Roman stoic philosopher and member of the Iroquois Confederacy.*

"Every man prefers belief to the exercise of judgment."

"Religion is regarded by the common people as true, by the wise as false, and by rulers as useful." *(Compare* LUCRETIUS.*)*

Captain Sensible (1954–), *guitarist and founding member of the punk band The Damned.*

"How many times have religions of the world been damaged by some discovery or other only to move the goalposts and carry on as before as though nothing had happened?"

William Shakespeare (1564–1616). *According to* George Santayana's *essay "Absence of Religion in Shakespeare," Shakespeare's entire opus contains no more than half a dozen passages "that have so much as a religious sound": "For Shakespeare, in the matter of religion, the choice lay between Christianity and nothing. He chose nothing." Which is, after all, what the idiotic tale of life signifies.*

"In religion, what damned error but some sober brow will bless it, and approve it with a text, hiding the grossness with fair ornament."—*Bassanio in* The Merchant of Venice.

George Bernard Shaw (1856–1950), *Irish-born English playwright. Speculated about some sort of "life force" twaddle, but declared at an early age that he was "like* Shelley, *a socialist, an atheist, and a vegetarian." His first published writing was a letter to a newspaper mocking an evangelical revival; those who attend and are converted, he wrote, become "highly objectionable members of society." A vigorous defender of* Charles Bradlaugh *in his battle with Parliament, yet attacked Bradlaugh's organization, the National Secular Society, as militant to the point of fundamentalism.*[3]

"The fact that a believer is happier than a skeptic is no more to the point than the fact that a drunken man is happier than a sober one."

"No man ever believes that the Bible means what it says: He is always convinced that it says what he means."

"We have not lost faith, but we have transferred it from God to the medical profession."

"Why should we take advice on sex from the pope? If he knows anything about it, he shouldn't!"

"Martyrdom is the only way in which a man can become famous without ability."

Last words: "Well, it will be a new experience anyway."

Martin Sheen (born Ramón Gerardo Antonio Estévez, 1940–), *American actor and 43rd president of the United States. Adopted his stage name in honor of Catholic archbishop and theologian Fulton J. Sheen.*

"I'm one of those cliff-hanging Catholics. I don't believe in God, but I do belief that Mary was his mother." *(Compare GEORGE SANTAYANA.)*

Percy Bysshe Shelley (1792–1822), *English poet and polemicist. Expelled from Oxford for writing an essay titled "The Necessity of Atheism." Rejected Deism, saying there can be no middle ground between believing and disbelieving in God. Drowned at age 29 when his yacht capsized in the Mediterranean while on his way to meet writers whom BYRON had invited to start a new periodical,* The Liberal. *His second wife, Mary, the daughter of MARY WOLLSTONECRAFT, wrote* Frankenstein, *whose implication that not only God could create a man anticipated the bioethical issues posed by cloning and artifical intelligence—this 200 years before Herman Munster.*

Opening line of "The Necessity of Atheism": "There is no God."

"The plurality of worlds—the indefinite immensity of the universe—is a most awful subject of contemplation. He who rightly feels its mystery and grandeur is in no danger of seduction from the falsehoods of religious systems."

"That which is incapable of proof [a Designer] itself is no proof of anything else [design]. . . . We must prove design before we can infer a designer."

"If he is infinitely good, what reason should we have to fear him? If he is infinitely wise, why should we have doubts concerning our future? If he knows all, why warn him of our needs and fatigue him with our prayers? If he is everywhere, why erect temples to him? If he is just, why fear that he will punish the creatures that he has filled with weaknesses? . . . If he is reasonable, how can he be angry at the blind, to whom he has given the liberty of being unreasonable? . . . If he is inconceivable, why occupy ourselves with him? . . . *and If he has spoken, why is the world not convinced?"*

From the poem Queen Mab: "I was an infant when my mother went / To see an atheist burned. She took me there / . . . And as the culprit passed with dauntless mien, / Tempered disdain in his unfaltering eye, / Mixed with a quiet smile, shone calmly forth;. . . . 'Weep not, child!' cried my mother, 'for that man / Has said, There is no God.' / . . . There is no God! / Nature confirms the faith his death-groan sealed. / . . . The name of God / Has fenced about all crime with holiness, / Himself the creature of his worshippers. . . ."

Marian Noel Sherman (1892–1975), *Canadian physician.*
Went "from missionary doctor [in India, 1922–1934] to atheist with a mission," wrote Annie Laurie Gaylor in Women without Superstition.

"Religious people often accuse atheists of being arrogant and of placing ourselves in the position of God, but really it is the theist who has all the vanity. He can't stand to think that he will ever cease to exist. As FREUD said, Christianity is the most egotistical of the religions. It is based on the premise 'Jesus saves me.'"

"If you tell a child 'God made the world' he will usually ask 'Then who made God?' If we reply, as the catechism states, 'No one made God. He always was,' then why couldn't we just say that about the world in the first place?"

Ricky Sherman (c. 1982–), *American former seven-year-old.*
From the Santa Rosa, California Press Democrat, *1989:*

"Putting a different spin on Flag Day, a seven-year-old atheist Wednesday urged public-school students to refuse to recite the Pledge of Allegiance until the words 'under God' are excised. . . . 'When kids are forced to say, "under God," it makes them think that atheists are bad people,' Ricky Sherman said at a news conference. 'Atheists are good people,' he said. 'We just know that God is make-believe.'"

Michael Shermer (1954–), *American science writer and historian*
of science. Founder of the Skeptics Society; editor of its magazine the Skeptic;
monthly columnist for SCIENTIFIC AMERICAN *magazine. His books include*
Why People Believe Weird Things. *Former marathon bicycle racer and*
founder of the Race Across America. Former fundamentalist Christian.

> "David Koresh, L. Ron Hubbard, Joseph Smith, Jesus, Moses,
> what's the difference? They were all egomaniacal, delusional
> characters who developed fanatical followers who exaggerated
> their claims, mythologized their lives, and canonized their
> words."

> "The only reason Stalin and Hitler killed more people than the
> Inquisition is that Torquemada didn't have gas chambers and
> machine guns."

Lionel Shriver (born Margaret Ann Shriver, 1957–) *U.S.-born,*
British-based novelist and columnist. Changed her name at age 15 because
she thought men had an easier life. Winner in 2005 of the prestigious,
£30,000 Orange Prize for the best novel by a female author with or
without a male name or hard life. On the controversial cancellation in 2004
of a play by a Sikh playwright in Birmingham, U.K., because Sikhs found
it offensive (it depicted murder and sex abuse in a Sikh temple):

> "The Roman Catholic Bishop of Birmingham applauded the
> cancellation . . . intoning that 'with freedom of speech and
> artistic license must come responsibility.' . . . Apparently,
> whatever is sacred to you must also be sacred to me. . . . Respect
> is earned; it is not an entitlement. . . . If I proclaim on a street
> corner that a certain Japanese beetle in my back garden is the
> new Messiah, you are also within your rights to ridicule me as a
> fruitcake."

Michelangelo Signorile (1960–), *American gay rights*
activist, columnist, and radio host on Sirius Satellite Radio's OutQ channel.
Author of the seminal book Queer in America. *Pioneer outer whose belt*
notches include tycoon Malcolm Forbes, the son of right-wing, anti-gay-rights

activist Phyllis Schlafly, and NBC correspondent/former Dick Cheney press
secretary/former assistant defense secretary Pete Williams.

> "All the while that Afghanistan's ruling Taliban has been
> protecting Osama bin Laden, Italy has been harboring another
> omnipotent religious zealot, one who equally condemns us
> Western sinners and incites violence with his incendiary
> rhetoric. . . . Meet John Paul II, Christian fundamentalist
> extraordinaire and a man who inspires thugs across the globe
> who commit hate crimes against homosexuals." *(2001)*

Sarah Silverman *(1970–), American comedian. Says she's "almost*
positive there's no god."

> "Everybody blames the Jews for killing Christ, and then the Jews
> try to pass it off on the Romans. I'm one of the few people that
> believe it was the blacks."

Joe Simpson *(1960–), British mountain climber. His 1988 book*
Touching the Void *described his and Simon Yates's near-fatal attempt to*
climb a peak in the Peruvian Andes. On saving himself from falling into a
crevasse:

> "I was brought up as a devout Catholic. I had long since stopped
> believing in God. I always wondered, if things really hit the fan
> whether I would, under pressure, turn around and say a few Hail
> Marys and say, 'get me out of here.' It never once occurred to
> me. If I had even thought that was the way out, or some sort of
> solace, or it was the time to meet my maker and go to paradise,
> I would have just stopped still. Then I would have died." *To*
> *"foxhole conversion," add the term "crevasse conversion." (I once*
> *came* this close *to a system-crash-no-data-backup conversion.*
> *Whatever doesn't kill you makes you stronger.)*

Rev. Timothy F. Simpson, *American Presbyterian minister.*
Director of religious affairs for the liberal-left Christian Alliance for Progress.
A press conference announcing the group's formation in 2005 was held in

front of the Jacksonville, Florida First Baptist Church, whose pastor had referred to MUHAMMAD *as "a demon-possessed pedophile." (I'm not sure whose side I'm on here.) Simpson also opposes "extremist" Democrats— "the five percent or so who don't want any religious rhetoric at all" in politics. (Now I'm sure.) Says that's why "Democrats get their butts kicked. Because people in this country are believers." To the First Baptist folks:*

> "We believe that you all—through your affiliation with the Southern Baptist Convention, which has become almost a wholly owned subsidiary of the Republican Party—have abandoned the values of our founder, Jesus Christ. . . . We understand that the Gospel is calling us to do very different things than just hobnob with the wealthy and lay down moral cover fire for the invasion of Iraq."

Frank Sinatra (1915–1998), *American singer and actor, not the biochemist. His mother, an immigrant from Italy, was an illegal abortionist known as "Hatpin Dolly" and a populist political organizer. Inscribed on his tombstone: "The best is yet to come."*

> "When lip service to some mysterious deity permits bestiality on Wednesday and absolution on Sunday, cash me out."

Upton Sinclair (1878–1968), *American socialist author. The uproar caused by his 1906 novel* The Jungle, *which dealt with conditions in the U.S. meat packing industry, led to the passage of the Meat Inspection Act— although he had intended to dramatize the conditions of the workers, not of the meat. Established a socialist commune called Helicon Hall. Helped found the California chapter of the ACLU. Won the Democratic nomination for governor of California in 1934. In 1927,* The Nation *magazine called him the* FIELDING *and* DICKENS *of his time. Arthur Conan Doyle called him "the* ZOLA *of America."*

> "A long time ago—no man can recall how far back—the Wholesale Pickpockets made the discovery of the ease with which a man's pockets could be rifled while he was preoccupied with spiritual exercises, and they began offering prizes for the best essays in support of the practice."

Peter Singer (1946–), *Australian philosopher and animal rights champion. Professed atheist. Challenges the position that all human life is sacred; says it is morally acceptable to euthanize severely disabled infants. Princeton University, where he was a professor of bioethics, thought it prudent to give him a scanner to check his mail for bombs. Singer also attacks "speciesism," which he says stems from the Judeo-Christian teaching that humans, made in the image of God, "have dominion" over animals, and that animals, since they have no souls, have no rights. Founding member of the Great Ape Project, which seeks to persuade the United Nations to adopt a declaration awarding personhood to nonhuman great apes. (Driver's licenses? The vote? Library cards?) Named Australian Humanist of the Year in 2004. In a study involving hypothetical moral dilemmas:*

> "There were no statistically significant differences between subjects with or without religious backgrounds. . . . Like other psychological faculties of the mind, including language and mathematics, [it appears that] we are endowed with a moral faculty that guides our intuitive judgments of right and wrong. . . . It is our own nature, not God, that is the source of our morality."

Frederick Smith, *American humorist? Originator of BD, or Beaver Design, the theory that our universe was created by one or more beaverlike creatures. Also called NID—Non-Intelligent Design.*

> "Now, it seems, there may be something called a 'meta-verse'— a region 'outside' or 'before' or 'bigger' than our universe. . . . Naturally, no one really knows anything about this region, if it exists, because by definition, we cannot examine it. . . . There are some interesting experiments planned, such as attempting to watch particles in accelerators 'leave' the universe. But, leave it to the ID [intelligent design] crowd, they not only accept such a region, but they believe that it houses a conscious, sentient, super-intelligent being. Indeed, so intelligent, that it created our universe!" *The obvious question: Could a particle accelerator cause larger particles, such as religious fanatics, to "leave the universe"? And could this "metaverse" place, with suitable modifications, permanently and securely contain them?*

George H. Smith (1949–), *American libertarian thinker. Author of* Atheism, The Case against God *(1979) and* Atheism, Ayn Rand and Other Heresies *(1991). Insists it is sufficient and necessary to define an atheist as "a person who does not believe in the existence of God" rather than as one who believes God does not exist: Since an atheist need make no claims about God, it is up to the believer to prove her case.*

"God is not matter; neither is nonexistence. God does not have limitations; neither does nonexistence. God is not visible; neither is nonexistence. God cannot be described; neither can nonexistence."

"Those who regard the fundamentalist revival as a harmless return to religious values would do well to take a closer look. There is a deep, underlying revolt here—a rejection of the rationalism and humanism in modern society."

"As [the Christian in search of converts] sees the matter, there is no wrong way to become a Christian. . . . It does not matter why you believe, so long as you believe."

"Christianity must convince men that they *need* salvation. . . . Christianity has nothing to offer a happy man. . . . *Just as Christianity must destroy reason before it can introduce faith, so it must destroy happiness before it can introduce salvation."*

"Christianity cannot erase man's need for pleasure, nor can it eradicate the various sources of pleasure. What it can do, however . . . is to inculcate guilt in connection with pleasure."

"Christianity has succeeded in convincing many people that misery incurred through sacrifice is a mark of virtue. . . . One invests in this life, so to speak, and collects interest in the next. Fortunately for Christianity, the dead cannot return for a refund."

Huston Smith (1919–) *American religious studies scholar. Born to Methodist missionaries in China, where he spent his first 17 years. Participated in some of* TIMOTHY LEARY's *drug experiments while a philosophy professor at MIT. Practiced Vedanta Hinduism, Zen Buddhism, and Sufism for over ten years each before turning back to Christianity—forgetting that by trading one religion for another, you're just avoiding the issues underlying your addiction.*

"Scientism smuggles in two untenable points. Namely, that science is, if not the only reliable, then the most reliable [way of knowing]. And second, that the stuff that science deals with, matter, is the most fundamental stuff of the universe. Those are not scientific statements. There is nothing in the way of science to prove they're true. And truth to tell, they are both wrong. . . . No one has ever seen a thought. No one has ever seen a feeling. Yet our thoughts and feelings are where we primarily live our lives. . . . Discounting invisible realities is the modern mistake promoted by an intolerant secularism that says only empirical, scientific knowledge is valid." *It all depends what is meant by "knowing."*

Quentin Smith*, American philosopher at Western Michigan University. Such a prolific writer, he's been said to forget to submit completed articles and even books to his publishers. Chief editor for* Philo: The Journal of the Society of Humanist Philosophers, *and an editor for Prometheus Books, which specializes in secularist/humanist works. Philosophers and even physicists from around the world solicit his opinions on their theories. Religious leaders seized on the Big Bang as a Creation story (see* HAWKING); *in* Theism, Atheism, and Big Bang Cosmology *(1993), Smith and Christian apologist philosopher William Lane Craig argued opposite sides of the question of whether the Big Bang supports theism or atheism.*

"God does not exist if Big Bang cosmology, or some relevantly similar theory, is true. If this cosmology is true, our universe exists without cause and without explanation. . . . Now the theistically alleged human need for a reason for existence [is] unsatisfied. But I suggest that humans do or can possess a deeper level of experience than such anthropocentric despairs. We can forget about ourselves for a moment and open ourselves up to the startling impingement of reality itself. We can let ourselves become profoundly astonished by the fact that this universe exists at all." *(A fact that Smith doesn't take for granted, by the way.)*

Stevie Smith (born Florence Margaret Smith, 1902–1971), *British poet and novelist. Never married. Apparently very picky:*

"If I had been the Virgin Mary, I would have said 'No.'"

Susan Smith (1971–), *American murderer. In 1994, the South Carolina mom locked her two sons in her car and rolled it into a lake for the sake of a relationship with a wealthy man who didn't want kids. Told the police an African-American carjacker drove away with her sons still in the car. Later confessed. Spared the death penalty after her stepfather, prominent Republican and Christian Coalition leader Beverly Russell, testified that he had sexually molested her when she was a teenager and again in the months before the drowning. Internet personal ad she placed in 2003:*

> "I am a Christian and I enjoy attending church. I consider myself to be sensitive, caring, and kind-hearted. I'm currently serving a life sentence on the charge of murder."

Lee Smolin (1955–), *American physicist/cosmologist. Most famous for his work on quantum gravity and his "fecund universes" theory, also known as "cosmological natural selection," which applies principles of biological evolution to cosmology, arguing that universes have the capacity to "reproduce." Our universe was born out of a black hole in another universe. (But don't spend the rest of your life trying to track down your cosmological parents. . . . You don't even know if they're still alive.)*

> "A scientific cosmology can contain no residue of the idea that the world was constructed by some being who is not a part of it. . . . As there can, by definition, be nothing outside the universe, a scientific cosmology must be based on a conception that the universe made itself. This is possible because, since DARWIN, we know that structure and complexity can be self-organized. . . . without any need for a maker outside of the system." *(Also see STUART KAUFFAMAN)*

> *The metaphysical implications of his research?* "The whole show of the universe is so extraordinary that the absence of God is God enough."

Raymond Smullyan (1919–) *American mathematician/logician/ philosopher. Started out as a stage magician. Wrote the definitive reformulation of Gödel's theorem, as you know. Author of popular books on recreational mathematics and logic puzzles.*

> "It has always puzzled me that so many people have taken it for granted that God favors those who believe in him. Isn't it

possible that the actual God is a scientific God who has little patience with beliefs founded on faith rather than evidence?"

From "Is God a Taoist?" (which, changing a few letters, becomes "God is toast"), a long dialogue between God and a mortal: "That I [God] should have been conceived in the role of a moralist is one of the great tragedies of the human race. . . . it is inaccurate to speak of my role in the scheme of things. I am the scheme of things. . . . I am the process [of attaining enlightenment]. . . . I am not the cause of Cosmic Process, I am Cosmic Process itself. . . . Those who wish to think of the devil (although I wish they wouldn't!) might analogously define him as the unfortunate length of time the process takes."

Annika Sörenstam (1970–), *professional golfer. Ranks among the top players on the LPGA Tour. First woman to play in a PGA event since 1945 (the Colonial tournament in Fort Worth, Texas, 2003).*

"Jag tror på det goda budskapet som finns inom religionen. Men att det finns någon däruppe ovanför molnen som styr är jag tveksam till."[*]

Mira Sorvino (1967–) *American actress and political activist affiliated with Amnesty International. Graduated magna cum laude from Harvard. Speaks fluent Mandarin.*

"Why does it not say anywhere in the Bible that slavery is wrong? . . . How is it possible that it is not immoral to own another person? Why isn't that one of the Ten Commandments? 'Thou shalt not own another person.' You want to sit here and tell me that fornication is worse than owning someone?"

John Lancaster Spalding (1840–1916), *American Roman Catholic bishop. Appointed Archbishop of Nazareth, a titular post since 1187.*

"Few really believe. The most only believe that they believe or even make believe [that they believe]."

[*] "I believe in the good message that's found in religion. But I doubt there's someone up there above the clouds running the show." (In Swedish.)

Bernard Spilka, Ralph Hood, and Richard Gorsuch, *authors of* The Psychology of Religion, *a standard text.*

"Most studies show that conventional religion is not an effective force for moral behavior or against criminal activity."

Baruch Spinoza (1632–1677), *Dutch-Jewish rationalist philosopher who prepared the way for the Enlightenment. Taught that God has no personality because "God and nature are one and the same" (pantheism); the Bible's truth is only metaphorical and allegorical; the Jews might* not *be God's chosen people; there is no real or absolute good and evil, and no afterlife. Labeled atheist by Jews and Christians and excommunicated by the Jewish community. Lens crafter by trade. "All our modern philosophers, though often perhaps unconsciously, see through the glasses which Baruch Spinoza ground."—*HEINRICH HEINE. *His portrait adorned the Dutch 1000-gulden bill until the Euro ruined everything.*

"If a triangle could speak, it would say, that God is eminently triangular, while a circle would say that the divine nature is eminently circular."

John Shelby Spong (1931–), *retired Episcopal bishop of Newark, New Jersey; controversial liberal theologian. Author of such bestsellers as* Rescuing the Bible from Fundamentalism *and* Why Christianity Must Change or Die. *Calls himself a nontheist.*

"The Bible is an ancient book. . . . There is no other piece of literature written in that period of history [c. 1000 B.C.E.–135 C.E.] which people today still treat as a source of ultimate truth. A doctor or pharmacist practicing medicine or dispensing drugs in our time based on either the writings of Aristotle or the formulas of an ancient medicine man would be laughed at first, and then if this activity were not stopped immediately, they would be accused of malpractice, removed from their professions and even imprisoned. . . . A chemist, biologist, architect or astronomer who acted on the basis of the knowledge available in the time the Bible was written . . . would be considered ignorant at best, mentally ill at worst."

"If the resurrection of Jesus cannot be believed except by assenting to the fantastic descriptions included in the Gospels, then Christianity is doomed."

"When will we recognise that religion is always in the mind-control business? . . . Organised religion is cultic at its core, but seeks to keep this fact well concealed. It is revealed only when its authority is questioned, or when some group takes the neurotic aspects of religion to their natural conclusion. That is the final meaning of the Heaven's Gate community in San Diego" *(a cult whose 39 members committed suicide in 1997 so that that their souls could board a spaceship they believed was hiding behind the Comet Hale-Bopp).*

Walter P. Stacy *(1925–1951), former chief justice, North Carolina Supreme Court.*

"It would be almost unbelievable, if history did not record the tragic fact, that men have gone to war and cut each other's throats because they could not agree as to what was to become of them after their throats were cut."

Madame de Staël *(Anne Louise Germaine de Staël, 1766–1817), Swiss-French novelist; mistress of a famous Paris literary-artistic-intellectual salon.*

"When woman no longer finds herself acceptable to men, she turns to religion."

Elizabeth Cady Stanton *(1815–1902), the preeminent fighter for women's rights in U.S. history. Her father, a lawyer, got her admitted to a previously all-male academy and taught her law, which women were forbidden to practice. At an antislavery conference she attended in London, Stanton and the other American delegate, Lucretia Mott—being women, basically—were, at the behest of the clergy, barred from participating and could only watch from behind a curtain. Two suffragist stars were born. Founded the National Woman Suffrage Association. Died 18 years before the 19th Amendment, allowing women to vote, was ratified.*

"The Bible and the Church have been the greatest stumbling blocks in the way of women's emancipation. . . . Among the clergy we find our most violent enemies, those most opposed to any change in woman's position." *Not the fundies—they want to roll it back!*

Lincoln Steffens (1866–1936), *American muckraking journalist and radical. Specialized in investigating government corruption. Said on his return from the USSR in 1921, "I have been over into the future [usually misquoted as "I have seen the future"], and it works." His enthusiasm for communism lasted only a few years, except in the FBI's imagination.*

"Why is it that the less intelligence people have, the more spiritual they are? They seem to fill all the vacant, ignorant spaces in their heads with soul."

"It is no cynical joke, it is literally true, that the Christian churches would not recognize Christianity if they saw it."

Gertrude Stein (1874–1946), *American-Jewish-born Modernist writer and art patron. Artistically radical, socially liberal (and lesbian), politically conservative or worse—sided with the Nationalist side during the Spanish Civil War* and, at first, supported the collaborationist Vichy government.*

"There ain't no answer. There ain't going to be any answer. There never has been an answer. That's the answer."

Gloria Steinem (1934–), *American journalist and womens' rights activist. Founder and original publisher of* Ms. *magazine. Member of Democratic Socialists of America.*

"It's an incredible con job when you think of it, to believe something now in exchange for life after death. Even corporations with all their reward systems don't try to make it posthumous."

---

* The estimated 350,000–500,000 people executed under Franco's orders after the war included atheists as well as leftists of all shades. "I am ready to execute half of Spain to do away with the reds," said Il Caudillo.

Addressing protesters in New York City during a visit by Pope John Paul II, 1995: "We will live to see the day that St. Patrick's Cathedral is a child-care center and the pope is no longer a disgrace to the skirt that he has on."

1973: "By the year 2000, we will, I hope, raise our children to believe in human potential, not God." *Poverty, hunger, disease, and war will be eradicated by then, too.*

"God may be in the details, but the goddess is in the questions. Once we begin to ask them, there's no turning back."

Stendhal *(pen name of Henri-Marie Beyle, 1783–1842) French writer, wit, dandy,* boulevardier, flâneur, farceur, tombeur. *Not fully appreciated until the twentieth century. (Nor will I be.)*

"All religions are founded on the fear of the many and the cleverness of the few."

Pope Stephen V*, Pope from 885 to 891.*

"The Popes, like Jesus, are conceived by their mothers through the overshadowing of the Holy Ghost. All Popes are a certain species of man-gods . . . all powers in Heaven, as well as on earth, are given to them."

Sir Leslie Stephen *(1832–1904), English writer on history, religion, and philosophy, and a leading agnostic (a term coined by THOMAS HUXLEY but popularized by Stephen). Born into a family of prominent evangelicals. Ordained as an Anglican priest in order to become a fellow at Cambridge. His works included* Essays on Freethinking and Plain-speaking *and* An Agnostic's Apology and Other Essays. *Married William Thackeray's daughter. Father (by his second wife) of VIRGINIA WOOLF.*

"They [religionists] feel rather than know. The awe with which they regard the universe, the tender glow of reverence and love with which the bare sight of nature affects them, is to them the ultimate guarantee of their beliefs. Happy those who feel such emotions! Only, when they try to extract definite statements of fact from these impalpable sentiments, they should beware how far such statements are apt to come into terrible collision with reality."

Howard Stern (1954–), *American TV and radio shock jock; king of shamelessness and tastelessness. All the inhibitions of a Tourette's sufferer. But none of his other outrages compares with his remark that "man created God." Then again, has said he starts praying to God when he has so much as a cold. Ranked #7 in* Forbes *magazine's 2006 list of the world's most influential celebrities, whatever that means. (It means the world is in extremely deep trouble is what it means.)*

"Please with the God talk."

"I'm sickened by all religions. Religion has divided people. I don't think there's any difference between the pope wearing a large hat and parading around with a smoking purse and an African painting his face white and praying to a rock."

"Here's what happens when you die—you sit in a box and get eaten by worms. I guarantee you that when you die, nothing cool happens."

Ian Stewart (1945–), *British mathematician and author of such books as* Does God Play Dice? The New Mathematics of Chaos.

"Science is the best defence against believing what we want to."

Max Stirner (pen name of Johann Kaspar Schmidt, 1806–1856), *German philosopher. Helped lay the foundation for nihilism, existentialism, and individualistic anarchism, which is a lot to answer for. Actually, disavowed all "isms" except egoism. Proclaimed that one must rid oneself of all sacred truths and artifical concepts—ideologies, religions, moralities, even science—and recover one's "fundamental state of existence," be one's own "creator and creature," and do whatever the fuck one wants, as I understand it.*

"The Holy Spirit became in time the 'absolute idea.' . . . Concepts are to decide everywhere, concepts to regulate life, concepts to rule. This is the religious world [of our time] . . . and the real man, I, am compelled to live according to these conceptual laws. . . . Liberalism simply brought other concepts on the carpet: human instead of divine, political instead of ecclesiastical, 'scientific' instead of doctrinal. . . ."

"The thinker is distinguished from the believer only by believing much more than the latter. . . . The thinker has a thousand tenets of faith where the believer gets along with few."

J. Michael Straczynski (1954–), *American TV producer and writer. Creator of the sci-fi TV series* Babylon 5 *and* Crusade. *Claims to have read the Bible cover to cover twice before deciding to become an atheist. (Once usually does it. [See* ELLEN JOHNSON.*]) On religion sullying his TV series:*

"In looking at the world 250 years from now [*when* Babylon 5 *is set*], I have to say that people will still believe . . . and I must treat that with respect. . . . Science and religion are two sides of the same coin. . . . both are endeavors to understand who we are, how we got here, where we are going, and what we are here to do."

Andrew Sullivan (1963–), *British-born American journalist. Former editor of* The New Republic. *Self-described "South Park Republican"; gay, HIV positive, conservative, practicing Roman Catholic—not the most comfortable of positions. Critic of the ultraconservatism and homophobia of Pope Benedict XVI. Supported the invasion of Iraq; supported John Kerry in 2004, largely because of the botching of Iraq (which seemed inevitable enough to folks less bright than Sullivan). On the Islamist response to modernism:*

"The temptation of American and Western culture—indeed, the very allure of such culture—may well require a repression all the more brutal if it is to be overcome."

"There is little room in the fundamentalist psyche for a moderate accommodation. The very psychological dynamics that lead repressed homosexuals to be viciously homophobic or that entice sexually tempted preachers to inveigh against immorality are the same dynamics that lead vodka-drinking fundamentalists to steer planes into buildings. It is not designed to achieve anything, construct anything, or argue anything. It is a violent acting out of internal conflicts." *(Also see* ROBERT M. YOUNG.*)*

William "Billy" Sunday (1863–1935), *American Protestant evangelist. One of the first mass-media (radio) evangelists. Previously a pro baseball player—until, one Sunday. . . .*

"The rivers of America will run with blood before they take our holy, God-inspired Bible from the schools."

"America is not a country for a dissenter to live in."

Supreme Court of Wisconsin

"There is no such source and cause of strife, quarrel, fights, malignant opposition, persecution, and war, and all evil in the state, as religion. Let it once enter into our civil affairs, our government would soon be destroyed. Let it once enter our common schools, they would be destroyed. Those who made our Constitution saw this, and used the most apt and comprehensive language in it to prevent such a catastrophe."—*From* Weiss v. District Board, *1890, in which the Court determined that Bible reading in public schools is unconstitutional. Previously, the King James Bible was recommended as a textbook by the state superintendent of schools.*

D. (Daisetz) T. Suzuki (1870–1966), *Japanese Zen Buddhist scholar and author. Almost singlehandedly popularized Zen in the West. Described by Catholic mystic Thomas Merton as "no less remarkable a man" than* EINSTEIN *or* GANDHI. *Has been criticized, along with other Japanese Zenmeisters, for helping to justify Japanese militarism before and during World War II (see* BRIAN DAIZEN VICTORIA).

"Zen has no God to worship, no ceremonial rites to observe, no future abode to which the dead are destined, and, last of all, Zen has no soul whose welfare is to be looked after by somebody else and whose immortality is a matter of intense concern with some people.. . . . This does not mean that Zen denies the existence of God; neither denial nor affirmation concerns Zen." *Zen doesn't get worked up about very much.*

"I discovered that it is necessary, absolutely necessary, to believe in nothing. . . . No matter what god or doctrine you believe in, if you become attached to it, your belief will be based more or less on a self-centered idea."

Jimmy Swaggart (1935–), *Fundamentalist televangelist and inveterate, whore-fucking adulterer. His was a $150 million-a-year "ministry" until film of him taking prostitutes to motels and allegations of a ten-year extramarital affair cut into his business. Claimed that Oral Roberts, over the phone, "cast out the demons" responsible for his behavior. Three years later, police stopped him for driving on the wrong side of the road and found him with another puta. Warned his congregation afterward that "God says it's none of your business!"*

On gay marriage: "If one ever looks at me like that, I'm gonna kill him and tell God he died."

"The Supreme Court of the United States of America is an institution damned by God almighty."

HONORABLE EVANGELICAL MENTIONS:

Oral Roberts: *Told his audience God told him he must raise $8 million or God would "take him home." Employees alleged he spent ministry funds on clothes, jewelry, and private jet travel.*

Robert Tilton: *Pulled in $80 mill a year while prayer requests sent with financial donations were being thrown away unread.*

Benny Hinn, televangelist/faith healer: *Alleged to use only a small percentage of his "ministry's" tax-exempt revenues for charitable purposes. Allegedly—just an allegation, mind you—fails to heal.*

Jim Whittington, evangelist: *Spent 10 years in prison for money laundering, mail fraud, conspiracy, and interstate transportation of stolen property. (Also see JERRY FALWELL, PAT ROBERTSON, BILLY SUNDAY.)*

Jonathan Swift (1667–1745), *Anglo-Irish priest, satirist, and political pamphleteer, mostly in support of Irish causes. In "A Modest Proposal (For Preventing the Children of Poor People in Ireland from Being a Burden)," he modestly proposed turning the children of the poor into food for the English rich—and suggested recipes. This was deemed "bad taste" by pettifogging prigs (some of whom perhaps wanted to buy the recipes alone, without the polemic). As Jesus said, the prophet is without honor in his own land.*

> "We have just enough religion to make us hate but not enough religion to make us love one another."

> "It is useless to attempt to reason a man out of what he was never reasoned into."

Algernon Charles Swinburne (1837–1909), *English poet, libertine, reprobate, and roué. Son of an admiral. Raised as an Anglican. Educated at Eton. Sent to Oxford. All the upper-class advantages a boy could want. How did he show his gratitude? By filling his poetry with sadomasochism, homosexuality, and anti-Christian sentiments. Called God "the supreme evil," and (compare* PAUL ERDÖS*) "humanity, not Jesus, the crucified victim."*[3]

> "Man, with a child's pride . . . Made God in his likeness, and bowed / Him to worship the Maker he made."

> "The beast faith lives on its own dung."

> "Thou hast conquered, O pale Galilean; the world has gone grey from your breath."

Thomas Szasz (1920–), *Hungarian-born American psychiatrist and leading critic of psychiatry. Author of* The Myth of Mental Illness *and* The Manufacture of Madness: A Comparative Study of the Inquisition and the Mental Health Movement. *Cofounded an anti-psychiatric organization in 1969 with the Church of Scientology. Relax: he's an atheist.*

"If you talk to God, you are praying; if God talks to you, you have schizophrenia. If the dead talk to you, you are a spiritualist; if you talk to the dead, you are a schizophrenic." *(Szasz mocked schizophrenia as "the Sacred Symbol of Psychiatry.")*

"Doubt is to certainty as neurosis is to psychosis. The neurotic is in doubt and has fears about persons and things; the psychotic has convictions and makes claims about them. In short, the neurotic has problems, the psychotic [*believer* or *atheist?*] has solutions."

Dr H. Tabar, *letter-writer to the* Times *of London, November 2001. Self-described apostate from Islam.*

"The cowardice of our clerics in pushing their heads firmly in the sand, not confronting the misguided and the extremists amongst us, is an affront to all that I regard as holy. If they have not the courage to declare the Islamic suicide terrorists as apostates, then perhaps they would be good enough to declare me as one, for I would rather burn in the eternal flames of Hell than share a Paradise with the likes of them." *(Reprinted on www.secularislam.org.)*

Takasui, *eighteenth-century Japanese Zen master.*

"Only doubt more and more deeply . . . without aiming at anything or expecting anything . . . without intending to be enlightened and without even intending not to intend to be enlightened. . . . When you become, through and through, a great mass of doubt, there will come a moment, all of a sudden, at which you emerge into a transcendence called the Great Enlightenment. . . ." *But first you must become a great, quivering, loathsome mass of doubt. . . . Hey, doubt's great—just don't deify it, dude.*

James Taylor (1948–), *American singer-songwriter. Earned his high school diploma while committed to a psychiatric hospital for depression in the 1960s. Recovering heroin addict. A hero for doing it without getting religion, and for facing up squarely to his hair loss.*

"I'm not saying that it's not helpful to think of having a real handle on the universe, your own personal point of attachment. But . . . I think it's crazy. But it's an insanity that keeps us sane. You might call a lot of these [his] songs 'spirituals for agnostics.'"

"Twelve-step programs say an interesting thing: Either you have a god, or you are God and you don't want the job." *What about (c) Have no god, are no god; (d) Are God, love the job?*

Edwin Way Teale (1899–1980), *American naturalist/photographer/ author.*

"It is morally as bad not to care whether a thing is true or not, so long as it makes you feel good, as it is not to care how you got your money as long as you have got it."

Tecumseh (1768–1813), *Shawnee leader who united Native American tribes against the invader. "One of those uncommon geniuses which spring up occasionally to produce revolutions and overturn the established order of things," said President William Henry Harrison. Insisted that Indian land was owned in common by all tribes, and thus no land could be sold without agreement by all.*

"How can we have confidence in the white people? When Jesus Christ came upon the earth, you killed him, the son of your own God, you nailed him up!! You thought he was dead, but you were mistaken. And only after you thought you killed him did you worship him, and start killing those who would not worship him. What kind of people is this for us to trust?"

Howard M. Teeple, *American Bible scholar. Author of* The Historical Approach to the Bible *(1982),* How Did Christianity Really Begin? *(1992) and the autobiographical* I Started to Be A Minister *(1990). Edits the* REI *(Religion and Ethics Institute) Digest.*

> "Bibliolatry: A form of idolatry, resulting from the acceptance of the Bible as an error-free rendition of divine inspiration. So much authority is assigned to the Bible that it, in effect, becomes the object of worship."

Woolsey Teller, *American astronomer. Author of* The Atheism of Astronomy: A Refutation of the Theory That the Universe Is Governed By Intelligence *(1938) and* Essays of an Atheist *(1945). Once debunked the claim of a University of Chicago professor to have proved the plausibility of the biblical Jonah story by crawling into the gullet of a whale. (Whales, Teller showed, have quite small gullets—proving that the entire Bible is bunk.)*

> "From every domain of science there is a wealth of evidence which shows the blind urge and senseless activities of natural phenomena. . . . The stellar depths are silent as the grave to human misery and want. The vast abyss of space is both our womb and our tomb. . . ."

> "There is something really pathetic in the statement that the universe was made for man. . . . There are more than 300,000 million stars [in our galaxy] alone [whose total weight is] equal to about 270,000 million suns the size of our own. This is the raw material, the amazing cosmic 'batter,' from which our planetary system came. . . . It is like mixing a batter of dough as big as the sun to bake a single crumb of bread. A baker who worked on the basis of that much material as a means to an end would be considered a dolt. . . . No mentality above the level of an idiot would devise such madhouse 'schemes' as that of spinning billions of globes for amusement or of tossing them around aimlessly to prove itself intelligent."

Charles Templeton (1915–2001), *Canadian journalist, novelist, and politician. Author of* Farewell to God: My Reasons for Rejecting the Christian Faith *(1995). Former evangelist and close colleague of Billy Graham. Declared himself an agnostic in 1957. Was offered the leadership of Canada's Liberal Party in 1967 but declined. Open nonbeliever is offered political leadership, and refuses . . . it's America Through the Looking Glass, eh?*

> "If God's love encompasses the whole world and if everyone who does not believe in him will perish, then surely this question needs to be asked: When, after two thousand years, does God's plan kick in for the billion people he 'so loves' in China? Or for the 840 million in India? Or the millions in Japan, Afghanistan, Siberia, Egypt, Burma—and on and on?" *Well, it was supposed to start (or rather, end) in 2000, but got pushed back for some reason.*

Alfred Lord Tennyson (1809–1892), *British Poet Laureate for the last 42 years of his life. Leaned toward agnosticism and deistic pantheism, or was it pantheistic deism. Praised* GIORDANO BRUNO *and* SPINOZA *on his deathbed.*

> "There lives more faith in honest doubt, believe me, than in half the creeds."

Mother Teresa (Agnes Gonxha Bojaxhiu, 1910–1997), *Albanian Catholic nun, "Angel of Mercy," "Saint of the Gutter." Founded the Missionaries of Charity in India. Beatified by Pope John Paul II in 2003, putting her halfway to sainthood (see* MATTHEW PARRIS*). Nobel Peace Prize, 1979; Presidential Medal of Freedom, 1985 (the perfect Reaganite/Bushite saint: let religious groups look after the poor—and let there be poor for them to look after). According to critics like* CHRISTOPHER HITCHENS, *author of* The Missionary Position: Mother Teresa in Theory and Practice, *money that her mission ostensibly raised for aid to the poor was spent largely on religious proselytizing. Susan Shields, an ex-member of Saint Teresa's staff, said of the mission's tens of millions of dollars in charitable contributions: "Most of the money sat in our bank accounts." According to a German*

magazine report, the Protestant-aligned Assembly of God charity serves many more meals daily in Calcutta alone than all the Missionaries of Charity's centers combined.

> "I think it is very beautiful for the poor to accept their lot, to share it with the passion of Christ. I think the world is being much helped by the suffering of the poor people."

Randall Terry (1959–), *right-wing evangelical-Christian ex-used car salesman and founder of the violently antiabortion group Operation Rescue. To group members:*

> "Let a wave of intolerance wash over you . . . a wave of hatred. . . . Yes, hate is good. . . . Our goal is a Christian Nation . . . we have a biblical duty, we are called by God to conquer this country. We don't want equal time. We don't want pluralism. We want theocracy." DIOGENES! *We found you an honest man!*

Tertullian (155–230), *Roman Christian priest/theologian; "father of the Latin Church." Originated the principle that the church's very existence was proof of the truth of its teaching, so that in any challenge or dissent against it, the burden of proof lay with the dissenter.*

> "You [women] are the gateway of the devil. . . . Because of what you deserve, that is, death, even the Son of God had to die. . . . Women, you ought to go about clad in mourning and rays, your eyes filled with tears of remorse, to make us forget you have been mankind's destruction."

Henry David Thoreau (1817–1862), *American philosopher/ naturalist; apostle of simple living amid nature. Protégé and fellow Transcendentalist of RALPH WALDO EMERSON. Lifelong abolitionist. Author of* On the Duty of Civil Disobedience. *Didn't pay his poll taxes for six years because he opposed slavery and the Mexican-American War. Typical America-hating Massachusetts liberal, siding with the terrorists. His ideas on natural history anticipate modern environmentalism.* New York Tribune

editor Horace Greeley refused some of his articles because of "your defiant pantheism" (belief that God is nature, basically). Pronounced his name THOR-eau.

> "I did not see why the schoolmaster should be taxed to support the priest, and not the priest the schoolmaster."

> *On missionaries in Canada:* "[There is no point trying] to convert the Algonquins from their own superstitions to new ones."

> "Do not be too moral. You may cheat yourself out of much life so. Aim above morality. . . . .When you travel to the Celestial City [*Los Angeles?*], carry no letter of introduction. When you knock, ask to see God—none of his servants. . . ."

> *Asked, as he neared death, if he had made his peace with God:* "I did not know that we had ever quarreled."

Leo Tolstoy (1828–1910), *Russian novelist and social activist. Inherited wealth, land, and the title of count, but hated Russia's brutal inequality and lived simply. Detested organized religion and its temporal power and rejected the supernatural in all forms. Wrote books like* Critique of Dogmatic Theology, *for which he refused payment. (Used the money from most of his later works to help the downtrodden.) These books were banned and burned and he was excommunicated by the Russian Orthodox Church (which he had described as "an impenetrable forest of stupidity"[*]). Also denounced from the pulpits for organizing relief from the famine that the czar and the church denied existed.*

> "The teaching of the church is in theory a crafty and evil lie, and in practice a concoction of gross superstition and witchcraft. . . . The Christian churches and Christianity have nothing in common save in name: they are utterly hostile opposites."

> "To regard Christ as God, and to pray to him, are to my mind the greatest possible sacrilege."

> "I was taught the soldier's trade, that is, to resist evil by homicide; the army to which I belonged *[as an officer in the Crimean War]* was sent forth with a Christian benediction."

---

[*] Russia remains a major lumber producer.

"People today live without faith. . . . The wealthy, educated people, having freed themselves from the hypnotism of the Church, believe in nothing. They look upon all faiths as absurdities or as useful means of keeping the masses in bondage—no more."

"A peasant dies calmly because he is not a Christian. He performs the rituals as a matter of course, but his true religion is different. His religion is nature, with which he has lived."

Tom Tomorrow (Dan Perkins, 1961–), *American left-liberal editorial cartoonist. His comic strip* This Modern World *depicts an absurdist world populated by dogmatic, robotic conservatives, a world unlike any TV network we know. It appears in around 150 papers and several book collections. On the 2002 federal appeals court ruling that having students recite the Pledge of Allegiance with the phrase "under God" is an unconstitutional endorsement of religion:*

"Our basic civil liberties are in jeopardy, but we're going to be spending our time as a society arguing about whether or not schoolchildren should be forced to pay tribute to imaginary invisible beings who live in magical kingdoms in outer space somewhere."

Polly Toynbee (1946–), *liberal British journalist; columnist for the* Guardian. *Topped a poll of 100 British "opinion makers"; named the most-read columnist in the U.K. Honorary Associate of the National Secular Society and a Distinguished Supporter of the British Humanist Association. Nominated in 2003 as "Most Islamophobic Journalist of the Year" by Britain's Islamic Human Rights Commission for her criticism of Islam's treatment of women—despite consistently defending immigrants and asylum seekers and attacking anti-Muslim bigotry. Yeah. Granddaughter of historian Arnold Toynbee.*

"The pens sharpen—Islamophobia! No such thing. Primitive Middle Eastern religions (and most others) are much the same—Islam, Christianity and Judaism all define themselves through disgust for women's bodies. . . . Meanwhile the far left, forever thrilled by the whiff of cordite, has bizarrely decided to fellow-travel with primitive Islamic extremism as the best available anti-Americanism around. (Never mind their new friends' views on women, gays and democracy.)"

"Religion is not nice, it kills: it is toxic in the places where people really believe it. . . . It is there in the born-again Christian fundamentalism demanded of every U.S. politician. . . . It drives on the murderous Islamic jihadists. It makes mad the biblical land-grabbing Israeli settlers. It threatens nuclear nemesis between the Hindus and Muslims along the India-Pakistan border. It still hurls pipebombs on the Ulster streets. The Falun Gong are killed for it, extremist Sikhs die for it too. The Pope kills millions through his reckless spreading of AIDS. When absolute God-given righteousness beckons, blood flows and women are in chains."

"The only good religion is a moribund religion. . . . [Religion] only becomes civilised when it loses all temporal power in a multicultural, secular society. . . . Only when the faithful are weak are they tolerant and peaceful. . . . Only then [does religion] turn into a gentle talisman of cultural tradition, a mode of meditation with little literal belief in ancient miracles or long dead warlords."

"I interviewed [MOTHER TERESA] . . . and we argued about contraception. Couldn't she see the effects of her teaching on the Calcutta streets where babies were born to starve and die in misery? She said that every baby that takes a breath is another soul to the glory of God and that was all that mattered, the creation of souls. Suffering? We are all born to suffer."

"Of all the elements of Christianity, the most repugnant is the notion of the Christ who took our sins upon himself and sacrificed his body in agony to save our souls. Did we ask him to?"

Leon Trotsky (born Lev Davidovitch Bronstein, 1879–1940),

Russian revolutionary. Key organizer of the October 1917 Revolution. Afterward, foreign affairs commissar and founder and commander of the Red Army. Organized a protest against an unpopular teacher in 2nd grade. Trostky was the name on the passport he stole in 1902 to escape from exile in Siberia. Do you think he and Frida Kahlo really . . . you know . . . ? 1933:

"Not only in peasant homes, but also in city skyscrapers, there lives alongside the twentieth century the thirteenth. A hundred million people use electricity and still believe in the magic powers of signs and exorcisms. Aviators who pilot miraculous mechanisms created by man's genius wear amulets on their sweaters. What inexhaustible reserves they possess of darkness, ignorance and savagery!"

Lao Tse *or* Laozi, *sixth century* B.C.E. *Chinese philosopher: founder of Taoism; author of its seminal work, the* Tao Te Ching.

"If lightning is the anger of the gods, the gods are concerned mostly with trees."

Ivan Turgenev (1818–1883), *Russian playwright and novelist* (Fathers and Sons). *Nonreligious, unlike* TOLSTOY *and* DOSOYEVSKY, *who were also, essentially, Russian novelists. Close friend of* GUSTAVE FLAUBERT. *After he died his brain was weighed: 2 kilograms. Adult average: 1–1.5 kg. (Religious average: no data available.)*

"Whatever a man prays for, he prays for a miracle. Every prayer reduces itself to this: 'Great God, grant that twice two be not four.'" *(Compare that latter-day Turgenev,* EMO PHILLIPS.*)*

"I shall be very curious to see the man who has the courage to believe in nothing." *It does take courage. Atheist can't even get insurance for acts of God.*

Agnes Sligh Turnbull (1888–1982), *American novelist. Author of the bestsellers* The Bishop's Mantle *(1948) and* The Gown of Glory *(1952). Far from an atheist; but a character in her 1936 novel* The Rolling Years *says, nice and quotably:*

"Wasn't religion invented by man for a kind of solace? It's as though he said, 'I'll make me a nice comfortable garment to shut out the heat and the cold'; and then it ends by becoming a straitjacket."

Ted Turner (1938–), *a.k.a "the Mouth of the South": American media mogul; president of the American Association of Media Moguls. Not. Founder of the first 24-hour cable news channel, CNN, and, perhaps equally importantly, the Cartoon Network. America's largest private land owner—his holdings exceed Delaware and Rhode Island combined in size. Professed agnostic. The American Humanist Association's 1989 Humanist of the Year. Says he was once more born-again than Ralph Reed.*

On the 1997 mass suicide by 39 Heaven's Gate cult members who believed their souls would be transferred onto a spaceship hidden behind comet Hale-Bopp:

> "I mean, is it that much different from other religions that say you're going to heaven?"

On why he and Jane Fonda split up:

> "She just came home and said 'I've become a Christian.' Before that, she was not a religious person. That's a pretty big change for your wife of many years to tell you. That's a shock."

Mark Twain (Samuel Clemens, 1835–1910), *American writer/journalist/humorist. Author of the strongly antireligious* Letters from the Earth. *(Oh, and* Huckleberry Finn *and* Tom Sawyer.) *His voluminous antireligious writings have been collected in a volume titled* The Bible According to Mark Twain.

> "Faith is believing what you know ain't so."

> "Religion consists of a set of things which the average man thinks he believes and wishes he was certain of."

> "There is nothing more awe-inspiring than a miracle except the credulity that can take it at par."

> "A man is accepted into a church for what he believes and he is turned out for what he knows."

> "I cannot see how a man of any large degree of humorous perception can ever be religious."

"It ain't the parts of the Bible that I can't understand that bother me, it is the parts that I do understand."

"The Bible commanded that [witches] should not be allowed to live. Therefore the Church . . . gathered up its halters, thumbscrews, and firebrands, and set about its holy work in earnest. She worked hard at it night and day during nine centuries and imprisoned, tortured, hanged, and burned whole hordes and armies of witches. . . . Then it was discovered that there was no such thing as witches, and never had been. One does not know whether to laugh or to cry."

"More than two hundred death penalties are gone from the law books, but the text that authorized them remains."

"The altar cloth of one aeon is the doormat of the next."

"If Christ were here now there is one thing he would not be—a Christian."

"No church property is taxed, and so the infidel and the atheist and the man without religion are taxed to make up the deficit in the public income this caused."

"Most people can't bear to sit in church for an hour on Sundays. How are they supposed to live somewhere very similar to it for eternity?"

"Go to heaven for the climate, hell for the company."

"I do not fear death, in view of the fact that I had been dead for billions and billions of years before I was born, and had not suffered the slightest inconvenience from it."

SECULAR * INFIDEL * NONBELIEVER * HUMANIST * RATIONALIST * FREETHINKER * AGNOSTIC * GODLESS * HERETIC * ATHEIST

Miguel de Unamuno (1864–1936), *Spanish writer and philosopher; forerunner of existentialists like* SARTRE *and* CAMUS *in his literary themes—spiritual anguish, the pain provoked by the silence of God— but responded very differently to that silence. Described faith as a "vital lie." Believed the irrational desire for immortality is what makes us human. Several of his works remained on the Vatican's list of prohibited books until the 1960s, and are still frowned upon regularly by Rome's Prefect of Frowning. Opposed both the Republic and Franco. Arrested in 1936; "died" in prison.*

> "Faith is in its essence simply a matter of will, not of reason, and to believe is to wish to believe, and to believe in God is, before all and above all, to wish that there may be a God."

> "Science as a substitute for religion, and reason as a substitute for faith, have always fallen to pieces."

> "The greater part of our atheists are atheists from a kind of rage, rage at not being able to believe that there is a God. They are the personal enemies of God. They have invested Nothingness with substance and personality, and their No-God is an Anti-God."

Peter Ustinov (1921–2004), *British actor, writer, and dramatist. president of the World Federalist Movement, which advocates a world government, from 1991 until his death. Indian Prime Minister Indira Gandhi was walking to an interview with him when she was assassinated in 1984.*

"Beliefs are what divide people. Doubt unites them."

Paul Valéry (1871–1945), *French Symbolist poet.*

"God made everything out of nothing, but the nothingness shows through."

Carl Van Doren (1885–1950), *American scholar, literary critic, and Pulitzer Prize–winning biographer. Longtime professor of English at Columbia University. A "world federalist" (like* PETER USTINOV, *who lived just a half-page before him). Uncle of renowned quiz show cheater Charles Van Doren. From his 1926 essay/statement* Why I Am an Unbeliever:

"Belief, being first in the field, naturally took a positive term for itself and gave a negative term to unbelief. . . . [But] what they call unbelief, I call belief."

"Many believers, I am told, have the same doubts, and yet have the knack of putting their doubts to sleep. . . . Believers are moved by their desires to the extent of letting them rule not only their conduct but their thoughts. An unbeliever's desires have, apparently, less power over his reason."

"No god has satisfied his worshipers forever. . . . In the case of the god who still survives in the loyalty of men after centuries of scrutiny, it can always be noted that little besides his name has endured. His attributes will have been so revised that he is really another god."

"The unbelievers have, as I read history, done less harm to the world than the believers. They have not filled it with savage wars . . . with crusades or persecutions, with complacency or ignorance. They have, instead, done what they could to fill it with knowledge and beauty, with temperance and justice, with manners and laughter. . . . They have surely not been inferior to the believers in the fine art of minding their own affairs and so of enlarging the territories of peace."

"I might once have felt it prudent to keep silence, for I perceive that the race of men, while sheep in credulity, are wolves for conformity. . . ."

Vincent van Gogh (1853–1890), *Dutch painter. Passionately religious in his early twenties. Took a missionary post in a Belgian mining town, but chose to live so poorly—sleeping on straw in a squalid hut—that the church dismissed him for "undermining the dignity of the priesthood." His father was by this time making inquiries about having him committed to a lunatic asylum.*

"I can very well do without God both in my life and in my painting, but I cannot, suffering as I am, do without something which is greater than I am, which is my life, the power to create."

Jesse Ventura (born James Janos, 1951–), *a.k.a. "The Body": former governor of Minnesota (1999–2003), professional wrestler, sports commentator, actor, radio talk show host, mayor, Navy SEAL and, briefly, bodyguard for the Rolling Stones. Modeled his wrestling persona on "Superstar" Billy Graham—the wrestler, not the Jesus tag-team partner. As governor (Reform/Independence Party), described himself as fiscally conservative and socially liberal; supported gay rights, abortion rights, medicinal marijuana, and church-state sep. Vetoed a bill to promote recitation of the Pledge of Allegiance in public schools. Achieved the highest approval rating of any governor in U.S. history. His supporters' slogan: "My governor can beat up your governor." (Evangelicals': "My god can beat up your god." [Really. See WILLIAM BOYKIN.])*

"Organized religion is a sham and a crutch for weak-minded people who need strength in numbers. It tells people to go out and stick their noses in other people's business."

Brian Daizen Victoria (1939–), *American-born Australian Buddhist priest, peace activist, and Asian Studies professor. No holy wars*

have ever been waged in the name of Buddhism—right? Victoria's book Zen at War *described how, from 1868 to 1945, Zen Buddhist leaders and institutions lent spiritual support to Japanese militarism and aggression.*

"[Zen scholar and popularizer D. T. SUZUKI] emphasized the iron will of Zen that could be wedded to anarchism or fascism, communism or democracy, atheism or idealism or any political or economic dogmatism."

"[Buddhism] was one, if not the only, organization capable of offering effective resistance to [Japanese] state policy. . . . [But] through the end of the Pacific War no major Buddhist or Christian leader ever again spoke out in any organized way against government policies, either civilian or military, domestic or foreign."

Shaku Soen, Zen master and teacher of Suzuki: "War is not necessarily horrible. . . . these sacrifices are so many phoenixes consumed in the sacred fire of spirituality. . . ."

Zen master Harada Daiun Sogaku, 1939: "[If ordered to] march: tramp tramp, or shoot: bang, bang. This is the manifestation of the highest Wisdom . . . The unity of Zen and war of which I speak extends to the farthest reaches of the holy war [now under way]."

Gore Vidal (1925–), *American novelist, essayist, screenplay writer, and professed "born-again atheist." His more than 30 novels include* Julian, *about the apostate late Roman emperor, and* Live from Golgotha: the Gospel According to Gore Vidal. *True or false?:* Myra Breckinridge *(1968), about an ex-boy named Myron with a taste for raping straight men with a strap-on dildo, was enormously popular with the religious right.*

"[Monotheism is] the great unmentionable evil at the centre of our culture. . . . I regard monotheism as the greatest disaster ever to befall the human race. . . . From a barbaric bronze age text known as the Old Testament, three anti-human religions have evolved—Judaism, Christianity, and Islam. These are sky-god religions. . . . The sky-god is a jealous god, of course. He requires total obedience from everyone on earth. . . . Although the notion of one god may give comfort to those in need of a daddy, it reminds the rest of us that the totalitarian society is grounded upon the concept of God the father. . . . Ultimately, totalitarianism is the only politics that can truly serve the sky-god's purpose. . . . One God, one King, one Pope, one master in the factory, one father-leader in the family at home."

"Any religion based on a single . . . well, frenzied and virulent god, is not as useful to the human race as, say, Confucianism, which is not a religion but an ethical and educational system that has worked pretty well for twenty-five hundred years. . . . But like it or not, the Book [the Bible] is there; and because of it people die; and the world is in danger."

"More people have been killed in the name of Jesus Christ than any other name in the history of the world."

C. F. Volney (Constantin-François de Chassebœuf, Comte de Volney, 1757–1820), *French savant. Author of* The Ruins, *a typically Enlightenment polemic on politics and religion in the form of a discourse on the fall of ancient empires between two observers floating high above the earth. The book was inspired by Volney's travels in the Ottoman Empire (and perhaps by whatever he smoked there). On priesthoods of all kinds:*

"They everywhere attributed to themselves prerogatives and immunities, by means of which they lived exempt from the burdens of other classes. . . . They everywhere avoided the toils of the laborer, the dangers of the soldier, and the disappointments of the merchant. . . . Under the cloak of poverty, they found everywhere the secret of procuring wealth. . . . In the form of gifts and offerings they had established fixed and certain revenues exempt from charges. . . . They styled themselves the interpreters and mediators [of God], always aiming at the great object to govern for their own advantage. . . . and all this by carrying on the singular trade of selling words and gestures to credulous people, who purchase them as commodities of the greatest value."

Voltaire (Francois Marie Arouet, (1694–1778), *French writer/philosopher. Crusader against political and religious cruelty and injustice! Espoused Deism . . . but who knows? In his day, atheism was punishable by death. Ended many of his letters and pamphlets with the slogan:* 'Écrasez l'infâme!" *("Crush the infamous thing"—i.e., Christianity.)*

From the Philosophical Dictionary (1764): "Atheist: A name given by theologians to whoever refuses to believe in God in a form of which, in the emptiness of their infallible pates, they have resolved to present it to him."

"Atheism is the vice of a few intelligent people."

"Doubt is not an agreeable condition, but certainty is an absurd one."

"Reason is the most hurtful thing in the world. God only allows it to remain with those he intends to damn, and in his goodness takes it away from those he intends to save or render useful to the Church."

"On religion, many are destined to reason wrongly; others not to reason at all; and others to persecute those who do reason."

"Those who can make you believe absurdities can make you commit atrocities."

"What can we say to a man who tells you that he would rather obey God than men, and that therefore he is sure to go to heaven for butchering you? [*Uh . . . "Allahu Akbar"?*] Even the law is impotent against these attacks of rage; it is like reading a court decree to a raving maniac."

"Superstition sets the whole world in flames; philosophy extinguishes it."

"The first priest was the first rogue who met the first fool."

"Every sensible man, every honorable man, must hold the Christian sect in horror. But what shall we substitute in its place? you say. What? A ferocious animal has sucked the blood of my relatives. I tell you to rid yourselves of this beast, and you ask me what you shall put in its place?"

"Another century and there will not be a Bible on earth!"

Kurt Vonnegut (1922–), *American novelist-satirist. Once called himself a "Christ-worshipping agnostic." Succeeded* ISAAC ASIMOV *in what Vonnegut called the "totally functionless capacity" of president of the American Humanist Association in 1992. The University of Chicago anthropology department inexplicably rejected his graduate thesis on the*

similarities between Cubist painting and nineteenth-century Native American uprisings, but later accepted his novel Cat's Cradle *and awarded him the degree. Has said his study of anthropology "confirmed my atheism." Yes, discovering the range of human beliefs, each one more bizarre than the next and all mutually contradictory, has a way of doing that. Once said everything one needs to know about good and evil was in* DOS-TOYEVSKY'S Brothers Karamazov.

> "The acceptance of a creed, any creed, entitles the acceptor to membership in the sort of artificial extended family we call a congregation. It is a way to fight loneliness."

> "I believe that virtuous behavior is trivialized by carrot-and-stick schemes, such as promises of highly improbable rewards or punishments in an improbable afterlife."

> "What the Gospels actually said was: don't kill anyone until you are absolutely sure they aren't well connected."

Frayba Wakili, *Afghan refugee from the Taliban; University of Maryland student thanks to a scholarship from the Feminist Majority Foundation. September 2000:*

> "Imagine being a teacher in a country where it is a crime to teach girls to count. Imagine living in a country where a child could be killed for learning the alphabet and opening a book. Imagine being beaten for organizing an underground library that distributes books to girls. This is happening in Afghanistan every day." *Thank God we drove the Taliban out. Well, at least out of downtown Kabul. (Look, we couldn't finish the job because we had to go bring Shiite clerics to power in Iraq, okay?)*

Alice Walker (1944–), *African-American novelist. Self-described "renegade, outlaw, and pagan." Her novel* The Color Purple *won the Pulitzer Prize.*

> "When I found out I thought God was white, and a man, I lost interest."

> *Shug, in* Purple: "She say, Celie, tell the truth, have you ever found God in church? I never did. I just found a bunch of folks hoping for him to show."

Peter Walker, *American space physicist; researcher at Rice University; contractor for the U.S. Naval Research Laboratory's Space Science Division.*

"The supreme arrogance of religious thinking: that a carbon-based bag of mostly water on a speck of iron-silicate dust around a boring dwarf star in a minor galaxy in an underpopulated local group of galaxies in an unfashionable suburb of a supercluster would look up at the sky and declare, 'It was all made so that I could exist!'"

Alfred R. Wallace (1823–1913), *simultaneous discoverer, with his colleague DARWIN, of natural selection. A half-hour's procrastination here, a missed bus there, and today it would be Wallace's theory of evolution we'd be trying to banish from our school curricula.*

"Man's special creation is entirely unsupported by the facts, as well as in the highest degree improbable."

Horace Walpole (1717–1797), *British politician and writer. Son of Prime Minister Robert Walpole; cousin of Admiral Horatio Nelson. Longtime member of Parliament. Gay member of Parliament, in an era in which one could declare "I am gay" without serious consequences. Trendsetter who popularized the brooding Gothic style (a queer eye had he) in architecture and literature.*

"What can be more ridiculous than to suppose that Omnipotent Goodness and Wisdom created [a heaven] and will select the most virtuous of its creatures to sing His praises to all eternity? It is an idea that I should think could never have entered but into the head of a king, who might delight to have his courtiers sing birthday odes forever."

Jill Paton Walsh (1937–), *English novelist and children's writer. On the new British law against "incitement of religious hatred," 2005:*

"Writers of fiction of whom I am one, are particularly vulnerable to attack, because of a widespread inability to read fictional works fictionally [rather than] as a statement of the author's view."

"The defence of tolerance would entail inviting the adherents of exogenic [of foreign origin] religions who live among us to accept that they must tolerate the customs and legal systems that have made our society attractive to them, as the price of being tolerated themselves. It is a profound moral duty not to claim for oneself what one will not concede to others."

Ibn Warraq, *life-preserving pseudonym of an agnostic, secular humanist Islamic scholar and outspoken critic of Islam. Author of* Leaving Islam: Apostates Speak Out *and* Why I Am Not a Muslim.

"Let us face the truth. . . . Islam divides the world in two: *Dar-ul Harb* [land of war] and *Dar-ul Islam* [land of Islam]. *Dar-ul Harb* is the land of the infidels. Muslims are required to infiltrate those lands, proselytize, procreate until their numbers increase, and then start the war . . . impose Islam . . . and convert that land into *Dar-ul Islam.* . . . And when the ignorant among us read those hate-laden verses, they act on them and the result is September 11, human bombs in Israel, massacres in East Timor and Bangladesh, kidnappings and killings in the Philippines, slavery in the Sudan, honor killings in Pakistan and Jordan, torture in Iran, stoning and maiming in Afghanistan and Iran, violence in Algeria, terrorism in Palestine and misery and death in every Islamic country. . . . It is not the extremists who have misunderstood Islam. They do literally what the Qur'an asks them to do. It is we who misunderstand Islam."

Lemuel K. Washburn, *early twentieth-century American secularist activist. Author of* Is the Bible Worth Reading? And Other Essays *(1911). Editor of the first U.S. rationalist newspaper, the* Boston Investigator.

"Of all the great inventions and discoveries that go to make human life easier, happier, more rich and glorious, not one can be laid to the work of theology. These triumphs all belong to science. . . . If man had no knowledge except what he has got out of the Bible he would not know enough to make a shoe."

"A man with a creed has bought the coffin for his mind. . . . It does not require any mental exercise to believe."

"Every fact is backed up by the whole universe."

"Some day the world will become wise enough to confess that the priest is of no benefit to mankind."

"Prayer is like a pump in an empty well, it makes lots of noise, but brings no water."

"The church is a bank that is continually receiving deposits but never pays a dividend."

"Jesus said: 'Follow me.' But we decline; we had rather not. . . . We do not think Jesus was a man that a self-respecting person would like to follow. . . . The man who said: 'believe and be saved, believe not and be damned,' cannot have our admiration."

"Religion is no more the parent of morality than an incubator is the mother of a chicken." "Lots of men who would not associate with infidels for fear of contaminating their characters are not yet out of jail."

"The Unitarian walks with a cane, the Congregationalist, Methodist, Presbyterian and Baptist go with crutches, the Episcopalian has to be pushed about in an invalid's chair, while the Roman Catholic crawls on his hands and knees and is led around with a ring in his nose by a priest."

"I side with the Agnostic. . . . The difference between people is this: Some don't know, and some don't know that they don't know, and the rest won't admit that they don't know."

George Washington (1732–1799), *our Founding Deist, and the Father of Church-State Separation. Attended church but refused Communion all his life. Never once mentioned Christ in his extensive correspondence. Instructed that workmen hired for his Mount Vernon plantation could be Muslims, Jews, Christians of any sect, or Atheists, for all he cared. On his deathbed, he requested no ritual and uttered no prayer. Had a Freemason funeral service.*

"The United States of America should have a foundation free from the influence of clergy."

Dr. James Watson (1928–), *American geneticist/biophysicist.*
Nobel-winning codiscoverer (with FRANCIS CRICK) of the structure of the
DNA molecule. Outspoken atheist. Strong proponent of genetically modified
crops, holding that the benefits far outweigh any plausible environmental dan-
gers. Has argued that the "really stupid" bottom 10 percent of people, iden-
tified through genetic screening, should be aborted before birth. (A rather
drastic way to eliminate religion. Just kidding. It's not as drastic as all that.)

> "The biggest advantage to believing in God is you don't have to
> understand anything, no physics, no biology. . . . I wanted to
> understand."

> "I don't think we're here for anything, we're just products of evolu-
> tion. You can say 'Gee, your life must be pretty bleak if you don't
> think there's a purpose,' but I'm anticipating a good lunch."

Wendell Watters, *M.D., psychiatrist and professor emeritus in*
psychiatry, McMaster University, Ontario. His 1992 book In Deadly Doctrine: Health, Illness, and Christian God-Talk *argued that by impeding*
healthy development of self-esteem, sexuality, and social interaction and pro-
moting sexual ignorance, Christian upbringing often leads to antisocial
behavior, sexual dysfunction, anxiety, and even major psychiatric illness. Has
this notion that Christianity's opposition to birth control and encouragement
of large families regardless of parents' emotional or financial resources under-
mines human welfare.

> "In devising a code of sexual behavior that would guarantee the
> survival of the church, the early fathers left no stone unturned in
> their determination to convert the female uterus into a factory
> for turning out Christian babies. . . . [Therefore] special
> condemnation was reserved for masturbation. . . . Western
> Christian society is the only one in which masturbation was
> totally proscribed." *When it should be prescribed.*

Another reason to be an atheist, period, colon: "A humanist is an
atheist who cares."

Bill Watterson (1958–), *American cartoonist/writer. Author of the award-winning comic strip* Calvin & Hobbes *(1985–1995). Calvin, an intelligent, troublesome six-year-old who maintains that his misbehavior is outside of his control, a product of his environment and circumstances, is named after sixteenth-century theologian John Calvin, a believer in predestination. Hobbes, Calvin's stuffed tiger come to life, is named after philosopher* THOMAS HOBBES *and shares his "dim view of human nature," as Watterson put it. Calvin:*

> "No efficiency. No accountability. I tell you, Hobbes, it's a lousy way to run a universe."

> "It's hard to be religious when certain people are never incinerated by bolts of lightning."

Alan Watts (1915–1973), *English-born American philosopher best known as a popularizer of Zen Buddhism (with a dash of Tao, a hint of Hinduism, a crust of Christianity, a vestige of Vedanta, a lick of linguistics, a phleck of physics, a cinder of cybernetics, an atom of anthropology, and 20 milligrams of psychiatry). After meeting D. T.* SUZUKI *and writing* The Spirit of Zen *in the 1930s, he earned a master's degree in theology at an Episcopalian seminary and worked as a priest. An extramarital affair resulted in his leaving the ministry. Became a prominent figure on the San Francisco hippie scene and a guru to the 1960s z-z-zeneration.*

> "Fanatical believers in the Bible, the Koran and the Torah have fought one another for centuries without realizing that they belong to the same pestiferous club, that they have more in common than they have against one another. . . . A committed believer in the Koran trots out the same arguments for his point of view as a Southern Baptist . . . and neither can listen to reason."

> The true believer . . . if he is somewhat sophisticated, justifies and even glorifies his invincible stpidity as a 'leap of faith' or 'sacrifice of the intellect.' He quotes the TERTULLIAN Credo, *quia absurdum est,* 'I believe because it is absurd' as if Tertullian had said something profound. Such people are, quite literally, idiots—originally a Greek word meaning an individual so isolated that you can't communicate with him."

"Today [1973], especially in the United States, there is a taboo against admitting that there are enormous numbers of stupid and ignorant people. . . . Many people never grow up. They stay all their lives with a passionate need for eternal authority and guidance. . . . This attitude is not faith. It is pure idolatry. . . . Faith is an openness and trusting attitude to truth and reality, whatever it may turn out to be. . . . Belief is holding to a rock; faith is learning how to swim."

"The architecture and ritual of churches [was] based on royal or judicial courts. A monarch who rules by force sits . . . flanked by guards, and those who come to petition him for justice or to offer tribute must kneel or prostrate themselves. . . . Is this an appropriate image for the inconceivable energy that underlies the universe?"

"If you picture the universe as a monarchy, how can you believe that a republic is the best form of government, and so be a loyal citizen of the United States? It is thus that fundamentalists veer to the extreme right wing in politics, being of the personality type that demands strong external and paternalistic authority."

"Things are as they are. Looking out into it the universe at night, we make no comparisons between right and wrong stars, nor between well and badly arranged constellations."

"The Buddhists . . . are not strictly atheists but feel that the ultimate reality cannot be pictured in any way and, what is more, that *not* picturing it is a positive way of feeling it directly, beyond symbols and images. I have called this 'atheism in the name of God.' [This] is an abandonment of all religious beliefs, including atheism, which in practice is the stubbornly held idea that the world is a mindless mechanism."

Evelyn Waugh (1903–1966), *satirical, reactionary English novelist. Quit the church of England at age 27 and became a staunch Catholic who opposed any and all church modernization.* Time *magazine: Waugh waged "a wickedly hilarious yet fundamentally religious assault on a century that, in his opinion, had ripped up the nourishing taproot of tradition." Wrote that after leaving Oxford without a degree, he attempted suicide by*

swimming out to sea but turned back after being stung by jellyfish. Once said, "I despise all my seven children equally."

"It is a curious thing that every creed promises a paradise which would be absolutely uninhabitable for anyone of civilized taste."

"I have noticed again and again since I have been in the Church that lay interest in ecclesiastical matters is often a prelude to insanity."

Simone Weil (1909–1943), *French philosopher, teacher, factory worker, and leftist trade unionist. Jewish family; raised agnostic. Proficient in Ancient Greek at age 12. Declared herself a Bolshevik at age 10. In 1937–38, had a pair of back-to-back religious experiences which tinged or tainted her subsequent writings with mysticism. Flirted with Roman Catholicism. Oy, Weil (pronounced "vey").*

"An atheist may be simply one whose faith and love are concentrated on the impersonal aspects of God."

"God can only be present in the creation in the form of absence." *Obscure (adj.): from French* obscur.

Steven Weinberg (1933–), *American theoretical physicist; 1979 Nobel winner for discovering, by God's grace, the unity of the atom's electromagnetic force and weak force. Criticizes religious conservatism for standing in the way of scientific inquiry and religious liberalism "for reducing theology to vacuousness in attempting to reconcile religion with science."* [4] *Damned if you do, damned if you don't.*

"With or without [religion] you would have good people doing good things and evil people doing evil things. But for good people to do evil things, that takes religion."

"One of the great achievements of science has been, if not to make it impossible for intelligent people to be religious, then at least to make it possible for them not to be religious."

"Most scientists I know don't care enough about religion even to call themselves atheists."

"The more the universe seems comprehensible, the more it seems pointless. . . . The more we refine our understanding of God to make the concept plausible, the more it seems pointless."

Tom Weller, *author of* Science Made Stupid: How to Discomprehend the World Around Us, *winner of the 1986 Hugo Award for Best Non-Fiction Book. This parody of a high school science textbook includes a satirical account of the creationism vs. evolution debate and drawings by Weller of fictional prehistoric animals such as the duck-billed mastodon. His Web site offers* Five Sure-Fire Zucchini Recipes *and "supports The Coalition to Undermine Traditional Values."*

"Several thousand years ago, a small tribe of ignorant near-savages wrote various collections of myths, wild tales, lies, and gibberish. Over the centuries, these stories were embroidered, garbled, mutilated, and torn into small pieces that were then repeatedly shuffled. Finally, this material was badly translated into several languages successfully. The resultant text, creationists feel, is the best guide to this complex and technical subject." *[creation vs. evolution].*

H. G. (Herbert George) Wells (1866–1946), *English writer best known for his science fiction novels, including* The Time Machine *(1895),* The Island of Doctor Moreau *(1896), and* The War of the Worlds *(1898). His 1914 novel* The World Set Free, *which prophesied (and named) "atomic bombs," inspired physicist Leó Szilárd's theory of the nuclear chain reaction and thus became self-fulfilling. Outspoken socialist; his name was high on an SS list of persons marked for immediate liquidation upon the Nazi conquest of England. Advocate of a world-state. Believed average citizens were too ill-informed to be entrusted with the vote. His last book suggested that the replacement of humanity by another species might not be a bad idea. (Okay, that's elitism.)*

"Indeed Christianity passes. Passes—it has gone! It has littered the beaches of life with churches, cathedrals, shrines and crucifixes, prejudices and intolerances, like the sea urchin and starfish and empty shells and lumps of stinging jelly upon the sands after a tide. . . . And it has left a multitude of little wriggling theologians and confessors and apologists hopping and burrowing in the warm nutritious sand. But in the hearts of living men, what remains of it now? . . . Phrases. Sentiments. Habits."

"Moral indignation: jealousy with a halo."

Alfred North Whitehead (1861–1947), *British mathematician and philosopher. Coauthor with* BERTRAND RUSSELL *of the magisterial* Principia Mathematica, *which an estimated three people (including the authors) have read in its entirety. Father of "process theology," which envisions God as changeable and nonomnipotent—as merely influencing the exercise of free will. (They'll perfrom any gymnastics to keep this Guy alive, won't they? They fail to grasp the secret of fundamentalism's success: No religious compromises with science and rationalism are sustainable; it has to be all or nothing.)*

"God is the ultimate limitation, and his existence is the ultimate irrationality. . . . No reason can be given for the nature of God because that nature is the ground of rationality."

"The total absence of humor in the Bible is one of the most singular things in all literature." *In stark contrast to the madcap mirth of the Koran.*

Walt Whitman (1819–1892), *America's most influential poet. Historians still can't decide whether his African-American lover down in N'orleans was a he or, as Whitman claimed, a she, and frankly, Americans are beginning to lose patience.*

"I have said that the soul is not more than the body / And I say that the body is not more than the soul, / And nothing, not God, is greater to one than one's self is. . . ."

"I think I could turn and live with animals / They are so placid and self-contain'd . . . / They do not lie awake at night in the dark and weep for their sins / They do not make me sick discussing their duty to God. . . . Not one kneels to another, nor to his kind that lived thousands of years ago. . . ."

"And I say to mankind, Be not curious about God. / For I, who am curious about each, am not curious / About God—I hear and behold God in every object, / Yet understand God not in the least."

"Science, testing absolutely all thoughts, all works, has already burst well upon the world—a sun, mounting, most illuminating, most glorious, surely never again to set. But against it, deeply entrench'd . . . the fossil theology of the mythic-materialistic, superstitious, untaught and credulous fable-loving, primitive ages of humanity."

Elie Wiesel (1928–), *Romanian-Jewish-American novelist and philosopher. Nobel Peace Prize winner, 1986. Taken to Auschwitz at age 15 or 16. Wrote in his memoir* Night *that he lost his faith on his first night there. His play* The Trial of God *recounted the true story of a group of rabbis in Auschwitz who put God on trial, pronounced a guilty verdict, then announced it was time for evening prayers.*

"Never should I forget that nocturnal silence which deprived me, for all eternity, of the desire to live. . . . Never shall I forget these moments which murdered God and my soul and turned my dreams to dust."

To an inmate who asked, "Where is God now" when the Gestapo hanged a small boy who, because he was too light, took a half hour to die: "Where is He? Here He is—He is hanging here on this gallows."

"It seemed as impossible to conceive of Auschwitz with God as to conceive of Auschwitz without God. The tragedy of the believer is much greater than the tragedy of the nonbeliever."

Oscar Wilde (1854–1900), *Anglo-Irish poet, novelist, and hugely successful playwright until his legal troubles began, ending with his conviction and imprisonment for "gross indecency." His first (female) love became smitten with/bitten by Bram Stoker and married him instead. When Wilde's play* Salomé *was refused a performance license because it contained biblical*

characters, he contemplated becoming a French citizen. Alas, it isn't true that his last words were, "My wallpaper and I are fighting a duel to the death. One or other of us has got to go"—he said that weeks before he died.

"When I think of all the harm the Bible has done, I despair of ever writing anything to equal it."

"To believe is very dull. To doubt is intensely engrossing."

"The worst vice of the fanatic is his sincerity."

How well Wilde understood Mel Gibstianity—or as he called it, "Medievalism, with its saints and martyrs, its love of self-torture, its wild passion for wounding itself, its gashing with knives, and its whipping with rods—Medievalism is the real Christianity, and the medieval Christ is the real Christ."

"Man can believe the impossible, but can never believe the improbable."

"I think that God, in creating men, somewhat overestimated his ability."

Thomas "Tennessee" Williams (1911–1983), *American playwright. Beaten up in Key West in 1979 by five teenage boys—either disgruntled theatergoers or participants in the spate of antigay violence that occurred after a local Baptist minister ran an antihomosexuality newspaper ad.*

Rev. T. Lawrence Shannon, the desperate, alcoholic, all-but-defrocked priest in The Night of the Iguana *who is now "collecting evidence" against God:*

"Look here, I said, I shouted, I'm tired of conducting services in praise and worship of a senile delinquent—yeah, that's what I said. . . . All your Western theologies, the whole mythology of them, are based on the concept of God as a *senile delinquent*. . . ."

Garry Wills (1934–), *American Catholic scholar, historian and author. Pulitzer winner, 1993. His* Nixon Agonistes—*which the* New York Times *said "reads like a combination of* H. L. MENCKEN, JOHN LOCKE *and* ALBERT CAMUS"—*landed him on Nixon's enemies list.*

"Can a people that believes more fervently in the Virgin Birth than in evolution still be called an Enlightened nation? . . . The secular states of modern Europe do not understand the fundamentalism of the American electorate. . . . In fact, we now resemble those nations less than we do our putative enemies. Where else do we find fundamentalist zeal, a rage at secularity, religious intolerance, fear of and hatred for modernity? . . . We find it in the Muslim world. . . . Americans wonder that the rest of the world thinks us so dangerous, so single-minded, so impervious to international appeals. They fear jihad, no matter whose zeal is being expressed."

Edmund Wilson (1895–1972), *preeminent American literary critic of the twentiethth century. Helped popularize* HEMINGWAY, *Dos Passos, Faulkner,* NABOKOV, *and his close friend* F. SCOTT FITZGERALD. *One scholar wrote of Wilson's "inability to apprehend the religious impulse," adding that "his atheism could be expressed with such valor and briskness that the most devout might be beguiled."*

"[Religions] are, of course, not merely impostures . . . not even mere legends and myths. . . . The resurrection of Adonis or Jesus serves not merely to celebrate the coming of spring [but to revive] morale in the celebrants. In the same way, the sacred dances performed by the American Indians may not be accompanied or followed by the rain that the celebrants invoke, but they help to keep up the tribe's spirits. . . . The power of prayer is real: when the Arab repeats his ritual . . . when the Protestant Christian appeals to his God, he is rallying his own moral forces."

"The word *God* is now archaic, and it ought to be dropped by those who don't need it for moral support. The word has the disadvantage of having meant already far too many things in too many ages of history and to too many kinds of people . . ."

Ending an audience with Pope Pius XII at the Vatican in 1945:

"Let's get out of here, for God's sake."

Edward O. Wilson (1929–), *American entomologist and evolutionary biologist. The father of sociobiology, which studies animal and human behavior as products of evolution. An atheist, "he says we need to invest some of the passion now reserved for traditional religion into caring for our environment."*[12]

"To understand biological human nature in depth is to drain the fever swamps of religious and blank-slate dogma."

"[DARWIN] did not abandon Abrahamic and other religious dogmas because of his discovery of evolution by natural selection, as one might reasonably suppose. The reverse occurred. The shedding of blind faith gave him the intellectual fearlessness to explore human evolution wherever logic and evidence took him. . . . Thus was born scientific humanism, the only worldview compatible with science's growing knowledge of the real world and the laws of nature."

"So, will science and religion find common ground, or at least agree to divide the fundamentals into mutually exclusive domains? A great many well-meaning scholars believe that such rapprochement is both possible and desirable [see STEPHEN JAY GOULD]. A few disagree, and I am one of them. I think Darwin would have held to the same position. The battle line is, as it has ever been, in biology. The inexorable growth of this science continues to widen, not to close, the tectonic gap between science and faith-based religion."

Robert Anton Wilson ("RAW," 1932–), *American writer/philosopher/cult figure. Best known for* The Illuminatus! Trilogy *(1975), a humorous examination of American conspiracy paranoia mixing fact and fiction as part of what Wilson called "Operation Mindfuck." "OM" is of course a practice of Discordianism, a "chaos-based religion" described by followers as "a religion disguised as a joke disguised as a religion." Also into Sufism, Futurology, "Quantum Psychology," and, yes, Neuro Linguistic Programming. Wilson and* TIMOTHY LEARY *held forth jointly about the "eight circuit model of consciousness," "reality tunnels," Space Migration, Intelligence Increase, and Life Extension. (All this must sound better if you're*

stoned.) Calls himself a "mystic agnostic" or "Model Agnostic"; describes his approach as 'Maybe Logic.'"

"[Model Agnosticism] consists of never regarding any model or map of Universe with total 100 percent belief or total 100 percent denial. . . . [Polish semanticist Alfred] Korzybski suggested dozens of reforms in our speech and our writings, most of which I try to follow. One of them is if people said 'maybe' more often, the world would suddenly become stark, staring sane. Can you see JERRY FALWELL saying: 'Maybe God hates gay people. Maybe Jesus is the son of God'? Every muezzin in Islam resounding at night in booming voices: 'There is no God except maybe Allah' . . . ? Think about how sane the world would become after a while."

"The Bible tells us to be like God, and then on page after page it describes God as a mass murderer. This may be the single most important key to the political behavior of Western Civilization."

Woodrow Wilson (1856–1924), *twenty-eighth U.S. president (1913–1921). Democrat—as this quote indicates:*

"Of course like every other man of intelligence and education I do believe in organic evolution. It surprises me that at this late date such questions should be raised."

Sherwin T. Wine (1928–), *American rabbi and founder in 1963 of the Society for Humanistic Judaism, which espouses cultural Judaism, Judaism without God. Describes his position as "ignosticism"—"finding the question of God's existence meaningless because it has no verifiable consequences." A bar/bat mitzvah at his Birmingham Temple in suburban Detroit consists of delivering a paper to the congregation on some Jewish hero or other.*

"One of the signs of personal strength is that we take blame for what we do wrong. The other sign is that we take credit for what we do right. We do not alienate our power by assigning it to someone else. . . . Strong people are comfortable in recognizing their own power . . . nor do they call their power 'a higher power.'"

Marlene Winell, *American psychologist. Specialist in recovery from religious indoctrination. Author of* Leaving the Fold: A Guide for Former Fundamentalists and Others Leaving Their Religion *(1993). Yes, first-hand experience.*

> "In the fundamentalist view, unbelievers have only two relevant attributes: They are potential converts and sources of temptation. As objects of evangelism, they are called 'crops to be harvested,' 'sheep to be found,' and 'fish to be netted.' Because of the danger of worldly influence (much like a contagious disease) . . . contacts must be superficial, geared toward evangelism only, and cut short if there is not a positive response. Since Christians are already full of truth, there is no need for them to listen, nothing for them to learn, and much for them to lose by admitting alternative views into their consciousness."

Alan Wolfe, *American political scientist; professor at Boston College; director of the Boisi Center for Religion and American Public Life at Boston College. Frequent contributor to the* New Republic, New York Times, Washington Post, Atlantic Monthly. *Author of* The Transformation of American Religion: How We Actually Practice Our Faith *(2003). Has conducted programs under U.S. State Department auspices to bring Muslim scholars to the United States to learn about church-state separation. (In the United States?)*

> "Evangelicalism's popularity is due as much to its populistic and democratic urges—its determination to find out exactly what believers want and to offer it to them—as it is to certainties of the faith. . . . More attention is paid to finding plenty of free parking and babysitting than to the proper interpretation of passages of Scripture."

Mary Wollstonecraft (1759–1797), *English author and pioneer feminist. Decided, after an arduous youth, to become "the first of a new genus"—a female intellectual. Mother of MARY SHELLEY.*

"[Clergy are] idle vermin who two or three times a day perform in the most slovenly manner a service which they think useless, but call their duty."

Virginia Woolf (1882–1941), *English novelist and essayist. Feminist? Deplored the term as connoting a narrow focus on women's issues. Preferred to be known as a "humanist." Daughter of famous agnostic* LESLIE STEPHEN. *Raised in a home in which people like Henry James and* GEORGE ELIOT *were always dropping by without calling first.*

"I read the book of Job last night—I don't think God comes well out of it."

"It is far harder to kill a phantom than a reality."

"Some people go to priests; others to poetry; I to my friends."

"To look life in the face, always, to look life in the face, and to know it for what it is . . . at last, to love it for what it is, and then to put it away."

Frank Lloyd Wright (1869–1959), *American architect. Brought up Unitarian. Broke the First Commandment of successful architectural practice: Run off not with thy client's wife. The scandal killed his U.S. career for a couple of decades.*

"Ugliness is a sin."

"I believe in God, only I spell it Nature." *His "organic" architecture itself approaches nature worship.*

Steven Wright (1955–), *unsmiling, Bozo-haired American comedian.*

"I was driving alone one day and I saw a hitchhiker with a sign saying Heaven. So I hit him."

X-Y-Z

NONBELIEVER * HUMANIST * RATIONALIST * FREETHINKER * AGNOSTIC * GODLESS * HERETIC * ATHEIST * SECULAR * INFIDEL *

Xenophanes (570–480 b.c.e.), *Greek philosopher.*

"Men imagine gods to be born and to have clothes and voice and body, like themselves. . . . If oxen, lions, and horses had hands and could make fashion of art, they would fashion gods in their own images. . . . The Ethiopians make their gods black and snub-nosed; the Thracians say theirs have blue eyes and red hair. . . ."

Cathy Young (born Ekaterina Jung, 1963–), *Russian-born American journalist. Contributing editor/columnist at the libertarian* Reason *magazine;* Boston Globe *columnist; contributor to the* New York Times, Washington Post, New Republic, *etc., etc. Research associate at the conservative-libertarian Cato Institute.*

"After September 11, some credited God with ensuring that there were far fewer people than usual both in the hijacked planes and in the targeted buildings. You'd think that God could have simply tipped off the FBI."

Edward Young (1683–1765), *British poet and dramatist.*

"By night an atheist half believes in a God."

Robert M. Young, *U.S.-born, London-based psychoanalytic psychotherapist; coeditor of the Web site human-nature.com. Walks the fine line between recognizing the complexities of, and shared responsibilities for, religious extremism and blaming the United States and the West for September 11.*

"I think the widest generalization embracing the emotional basis of fundamentalism is the fear of annihilation of a way of life. This unites the impoverished and immiserated suicide bombers in Israel with the Saudis who adhere to the traditional Wahabi Muslim sectarianism from which al-Qaeda draws nourishment. This accounts for the presence of the relatively well-off among the perpetrators of the attacks of 9-11. The annals of those who oppose established orders in the name of a rigid faith are full of people who were economically prosperous but—arguably—psychologically threatened and damaged. . . . Freud has a motto on the frontispiece of his masterpiece *The Interpretation of Dreams* (1900) that I think sums up how fundamentalism links to terrorism at both a social and individual level: 'If I cannot bend the higher powers, I will stir up the lower depths,' it goes."

"It is tempting to defend oneself from feeling so abject by becoming, in fantasy, the opposite and attaining a position of complete self-sufficiency or certainty. Osama Bin Laden's father died when he was still a boy; his mother, not one of the father's main wives, was looked down upon. The young Hitler had an unhappy childhood and was a failed painter. 'I am nobody and am sure of nothing' becomes 'I am powerful and sure about everything: it is in the book.' If fundamentalists were really sure they would not have to be so intolerant."

Henny Youngman (1906–1998), *Liverpool-born American-Jewish comedian. "Take my wife . . . please!" originated as a simple request to an usher to show his wife to her seat. The usher laughed. The rest is history.*

"I wanted to become an atheist but I gave up. They have no holidays." (*This must change. Darwin Day? Newtonmas? Copernicalia? God Freeday?*)

Lin Yutang (1895–1976), *Chinese-American writer, translator of classic Chinese texts, and inventor. Taught English literature at Peking University. Invented and patented a Chinese typewriter and, more importantly, a toothbrush that dispensed its own toothpaste. Frequently nominated for the Nobel Prize in Literature.* My Country and My People *(1935) and* The Importance of Living *(1937), written in English, made him famous.*

> "The world of pagan belief is a simpler belief. . . . It does not encourage men to do, for instance, a simple act of charity by dragging in a series of hypothetical postulates—sin, redemption, the cross, laying up treasure in heaven, mutual obligation among men on account of a third-party relationship in heaven—all so unnecessarily complicated and roundabout, and none capable of direct proof."

> "In the West, the insane are so many that they are put in an asylum *[or, when there is overflow, in places like Congress, the White House, and the Christian Broadcasting Network]*. In China the insane are so unusual that we worship them."

> "Such religion as there can be in modern life, every individual will have to salvage from the churches for himself."

Frank Zappa (1940–1993), *American avant-garde comic proto punk jazz Dadaist Surrealist rock composer-guitarist-singer-record producer and detester of labels. Accordingly, objects named in his honor include an asteroid, a gene, a fish, a jellyfish, an extinct mollusc, a spider, and a street in Berlin, Frank-Zappa-Straße. Former Czech President Václav Havel named him special ambassador to the West on trade, culture and tourism. Still more prestigious: Three of his songs made Ann Landers's list of the ten most obscene rock songs. If you recognize the title* Trout Mask Replica, *you are a Zappatista.*

> "The essence of Christianity is told us in the Garden of Eden story. The subtext is, All the suffering you have is because you wanted to find out what was going on. You could be in the Garden of Eden if you had just kept your fucking mouth shut and hadn't asked any questions. . . . 'Get smart and I'll fuck you over,' sayeth the Lord. Is this not an absolutely anti-intellectual religion?"

"America was founded by the refuse of the religious fanatics of England, these undesirable elements that came over on the Mayflower. Ignorant, religious fanatics who land here, abuse the Indians, and then go to bed with a board down the middle, you know, the bundling board, so they don't have sex. That's how we got started."

"I don't want to see any religious people in public office because they're working for another boss."

"The only difference between a cult and a religion is the amount of real estate they own."

"My best advice to anyone who wants to raise a happy, mentally healthy child is: KEEP HIM OR HER AS FAR AWAY FROM A CHURCH AS YOU CAN."

"There is no hell. There is only . . . FRANCE!!!"

Frank Zindler, *American science writer. Spokesperson for American Atheists, editor of their magazine and director of their publishing house.*

"Science is reductionistic in that it tries to explain the unknown in terms of the known. Contrariwise, religious explanations frequently explain the unknown in terms of the even less known—the old fallacy of *ignotum per ignotius*" [*"unknown by means of the more unknown"*].

Slavoj Žižek (1949–), *Slovenian sociologist/philosopher/cultural critic. Writes on popular culture, cyberspace, fundamentalism, globalization, human rights. . . . Studied Lacanian psychoanalysis in Paris, for Christ's sake. Can write about whatever the fuck he wants. In fact, wrote the text for an Abercrombie & Fitch catalogue, not to mention a three-part documentary for British TV in 2006:* The Pervert's Guide to Cinema. *Calls his work a "materialist theology": Believes a slight but crucial gap always remains between the Real of life (not to be equated with reality) and what materialist science can explain.*

"More than a century ago . . . DOSTOYEVSKY warned against the dangers of godless moral nihilism, arguing in essence that if

God doesn't exist, then everything is permitted. The French philosopher André Glucksmann even applied Dostoyevsky's critique of godless nihilism to 9/11, as the title of his book, *Dostoyevsky in Manhattan*, suggests. This argument couldn't have been more wrong: the lesson of today's terrorism is that if God exists, then everything, including blowing up thousands of innocent bystanders, is permitted—at least to those who claim to act directly on behalf of God, since, clearly, a direct link to God justifies the violation of any merely human constraints and considerations."

"When I do a good deed, I do so not with an eye toward gaining God's favor; I do it because if I did not, I could not look at myself in the mirror. A moral deed is by definition its own reward. DAVID HUME, a believer, made this point in a very poignant way, when he wrote that the only way to show true respect for God is to act morally while ignoring God's existence."

Émile Zola (1840–1902), *French novelist and (pardon the term) crusading liberal editorialist. For publicly j'accusing the French government of anti-Semitism in the Dreyfus Affair—the wrongful imprisonment of Jewish army captain Alfred Dreyfus for espionage—Zola was tried and convicted of libel. Fled to England. But his efforts led to the reopening of the case and to Dreyfus's eventual acquittal. It took the Roman Catholic daily* La Croix *100 years to apologize for its anti-Semitic editorials during* l'affaire.

"Has science ever retreated? No! It is Catholicism which has always retreated before her, and will always be forced to retreat."

"Civilization will not attain to its perfection until the last stone from the last church falls on the last priest!"

Sources

Main sources of biographical information and quotes:

1. Wikipedia.com
2. Jennifer Hecht, *Doubt: A History: The Great Doubters and Their Legacy of Innovation from Socrates and Jesus to Thomas Jefferson and Emily Dickinson*
3. James A. Haught, ed., *2000 Years of Disbelief*
4. CelebAtheists.com
5. Freedom From Religion Foundation (ffrf.org)
6. Internet Movie Database (imdb.com)
7. World Pantheism (pantheism.net)
8. Paul Johnson, *Intellectuals*
9. Jim Walker, "Thomas Jefferson on Christianity & Religion," Nobeliefs.com.
10. Karen Armstrong, *A History of God*
11. American Atheists (atheists.org)
12. Salon.com
13. AtheistAlliance.org
14. AtheistsUnited.org
15. David Boulton, "Who Needs Religion?" *New Internationalist* magazine, August 2004
16. Unitarian Universalist Historical Society (uua.org)

Main sources of quotes:

Adherents.com
Wayne Aiken's Atheist Fortune Cookie File
An Amateur's Guide to Heresy: Quotes for the Freethinker (starlingtech.com)
American Humanist Association
apatheticagnostic.com
atheistsunited.org/wordsofwisdom
The Brights Net

Eric's Quotations Related to Atheism" (edp.org)

Famous Atheists, Freethinkers, Deists and Agnostics (wonder-fulatheistsofcfl.org)

Sam Harris, *The End of Faith: Religion, Terror, and the Future of Reason*

HolySmoke.org

infidelguy.com

infidelityblog.org

Inquisitive Atheists (www.geocities.com/inquisitive79)

S. T. Joshi, ed., *Atheism: A Reader*

julianbaggini.com

Margaret Knight, *Humanist Anthology: From Confucius to Attenborough*

palmyra.demon.co.uk

positiveatheism.org

Reasoned Spirituality (reasoned.org)

Paul Rifkin, *The God Letters*

Secular Web

George Seldes, ed., *The Great Thoughts*

ACKNOWLEDGMENTS

My thanks to Shelley Hopkins for taking her cruel but ultimately merciful editorial machete to my impossibly overgrown manuscript, and in other ways helping make the thing between 1.6 and 2.4 times better than it would otherwise have been (as this very sentence suggests). Thanks also to Ruth Baldwin and Carl Bromley at Nation Books, and to my beloved and ever-supportive Renée.